North
Somerset
COUNCIL

D1421249

I NS
1 2 0319239 5

In Search of Steam 1962-68

To Jane

In Search of Steam 1962-68

Robert Adley

'And some will mourn in ashes, some coal-black
For the deposing of a rightful King . . .'

Richard II
Act V Scene ii

BLANDFORD PRESS
Poole Dorset

First published in the U.K. 1981, by Blandford Press
Link House, West Street, Poole, Dorset, BH15 1LL

Reprinted 1981
Reprinted in this edition 1982
Reprinted 1983

British Library Cataloguing in Publication Data

Adley, Robert
In search of steam 1962–68.
1. Locomotives – Great Britain – History – 20th century
I. Title
625.2'61'0941 TJ603.4.G7

ISBN 0 7137 1091 8

Typeset by Keyspools Ltd, Golborne, Lancs.

Produced in Great Britain by Shenval Marketing, Harlow

Contents

INTRODUCTION

Why do books have introductions? Is it to enable the author to explain or apologise, or is it for no reason other than tradition? If explanation be needed, then questions must be asked. Do *you* have to explain why you bought this book? Does the publisher have to explain why he decided to publish it? Do I have to explain why I wrote it? My explanation is straightforward – I love trains and railways. The publisher's explanation is that he hoped you would want to buy it.

I regard the steam engine as the most interesting, attractive and exciting of man's creations. Thus I photographed its declining years in Britain, between 1962–1968, purely for my own pleasure, in order to enjoy my memories once steam had been displaced on British Rail. My photographs determine the scope of this book.

Is that sufficient? Perhaps not. The physical domination of railway over landscape, particularly urban landscape, ensured that it could not be ignored. Much remains to remind us of the impact on our country wrought by the railway engineers, Stockport viaduct, as illustrated, was completed in 1842. Other than the smoke, the scene is not so different today. The men who built and ran the railways dominated the social environment, from engineer to navvy, from financier to signalman. Therefore my reason for writing this book is to combine creative nostalgia with recognition of the social and historic importance of, and interest in, the railway.

Compare the relationship of rail to landscape, with the relationship of road to landscape. If the railway dominated, it was domination without destruction. It attracted the eye rather than repelled it. Trails of steam and smoke in the Lune Gorge were a feature, not an eyesore. Joseph Locke was an engineer; he was an engineering artist, too.

Who built the M6? Who knows? Who cares? Whoever he was he damaged nature's creativity, not enhanced it. The fact that

Fig. I. Stockport Viaduct with the London & North Western Railway crossing the valley.

nobody 'built' the M6, or any motorway, merely illustrates the boring impersonality plus the damaging environmental scar of road on landscape. Railways are alive, are people: roads are dead things.

By word and illustration, the steam railway lives on.

Simultaneously with the passing of the steam era has come the rapid improvement of techniques for colour film manufacture and reproduction; at the same time, the craving for written and pictorial memorials to the steam railway has grown rather than diminished. Thus anyone with an unpublished collection of colour photographs of British working steam who was too shy or too lazy to publish them, or was cunning enough to delay their use until now, may be rewarded. The elapse of time between the end of British Rail steam in August 1968 and the present day is now encouraging railway photographers and authors to contemplate writing railway books to an untried formula. Thus I present you with my second railway book; combining as it does, full-colour photographs from my camera with anecdotes from my experience and aspects of railway history which have caught my imagination – the latter being illustrated by the former.

As railway enthusiasts are born, not made, so we have grown-up with our childhood heroes. Ransome Wallis and Casserley are names to be revered. They adorned the captions in the railway books upon which we were weaned!

As the steam era recedes into the past, the anticipated diminution of interest in steam railways has not occurred, and we can discern a changing pattern in the appearance of today's railway books. In presenting his book, *Roaming the West Coast Rails*, Derek Cross sought, as I am doing, to combine anecdotal, historical, evocative and photographic aspects of the steam railway into a book more lively than an historic treatise, yet more evocative and detailed than solely illustrative. I hope we both are right!

Searching for a title was a lengthy process. As few suitable adjectives remain unattached to the word *steam* among railway books published during the last twenty years, I have borrowed my theme from H V Morton, to whom, posthumously, I apologise.

Hours spent with Roget's *Thesaurus* produced an endless stream of adjectives; but whenever I thought I had found the answer, it was only to discover that someone had beaten me to it: or that my enthusiasm rarely lasted overnight. Then I thought I might copy Ivo Peters' *Somewhere Along the Line* by inventing a title without the word *steam*. I considered *line, track* and *rails* – it didn't click and the problem started to nag away at my mind.

Then Morton's *In Search of England* caught my eye as I rearranged books on my study bookshelves. One of the few books from my late father-in-law's collection to remain in the family, the copy has a map of England for endpapers.

As an Old Uppinghamian, my idle reference to places in the index caused me to check what he had to say about Rutland.

'I am the only person I have ever known who has been to Rutland. I admit that I have known men who have passed through Rutland in search of a fox, but I have never met a man who has deliberately set out to go to Rutland: and I do not suppose you have.'

Morton was wrong! But that did not bother me. *In Search of Steam* was conceived and birth was consummated when, amongst a job lot of books bought at a sale in Ringwood, I found *The Call of England* among a quantity of books which can most politely be described as 'mixed'.

The more I read of Morton, the more I realised that he had little time for railways, or steam. So, with passing thanks for the idea for the title of this book, we bid farewell to H V Morton, leaving him to his . . . .

'remarkable system of motor-coach services which now penetrates every part of the country which has thrown open to ordinary people regions which even after coming of the railway were remote and inaccessible.'

Any man who can compare favourably, the motor-coach to the train; or prefer the impact of roads on the environment to that of the railway, deserves no sympathy! Perhaps I can retaliate by hijacking his title theme.

The success of this book depends upon the accuracy of some of my assumptions. Let me list these assumptions. Railway enthusiasts like timetables, charts, railway engravings and, particularly, maps. For the inclusion of these items, I must offer thanks to the BBC Hulton Picture Library for the illustration of Stockport Viaduct, to the Ordnance Survey in Southampton, to British Rail, the Elton Collection, the Ironbridge Gorge Museum Trust and to Julia Elton, daughter of the late Sir Arthur Elton. Julia helped me locate many of the engravings and encouraged me to eradicate some of my more excessive bouts of steam sentimentality.

For such specific help and guidance I am truly grateful. Railway enthusiasts are almost impossible people with whom to live, to which my wife will surely testify. Jane has tolerated my eccentricities for nearly twenty years, and tells me when to stop being boring and obsessive. As a trained interior designer her knowledge of style, layout, theme and content tempers my enthusiasm with reality. She tries to keep me on the track of normality.

Jane's impact on my six years frantic journeying 'in search of steam' was more than geographic; although geography dictated my photography. Her parents lived in Burnham-on-Sea, the northernmost outpost of the Somerset & Dorset Joint Railway. Although the S & D is – was – perhaps the best loved, most individual and most photographed railway in Britain, I was unconscious of its proximity when first I visited Burnham.

Jane and I were married in August 1961 and by late 1962 she realised the need to propel me towards some form of activity that would get me off my backside at weekends and fill my lungs with fresh air. Jane's brother, Mickey Pople, lives in Bristol, and it was to Temple Meads that I made an early sortie with my camera. His two boys accompanied me to Barrow Road shed and my picture of them alongside No. 6919 *Misterton Hall* remains one of my favourite photographs. It is reproduced in this book for the first time.

Visits to my parents gave me an opportunity to go in search of steam around Brighton. From our home in Chiswick and later from Sunningdale, Jane and I radiated westwards to Somerset and southwards to Sussex. The GWR main line to Bristol and the West, and the Southern's main line to Bournemouth, Salisbury and the West, became my stamping grounds. All the engine sheds in the area became my weekend venues. Luckily my business life in the early sixties took me almost anywhere I wanted to go. As soon as I embarked on my search for steam, my business trips not surprisingly began to coincide with steam's remaining strongholds!

By 1965 politics had so intruded into my life that I was visiting Birkenhead almost every weekend (see Chapter 13 for the explanation). Thus my geographic axis was centred on Sunningdale, running north to Lancashire and Merseyside, west to Bristol and Somerset, and south to Sussex. The extremities of my photographic journeys took me north-east to Newcastle and Edinburgh, east to March, south-east to Tonbridge, west to Highbridge and to numerous points between. I was particularly fortunate with the sun in the Manchester area. Wherever I went, my Voigtlander Vito CLR went also, and wherever I was, the relevant Ordnance Survey map was with me. So too was my Ian Allan *Shed Directory* and 'abc', plus the inevitable thrice-folded pieces of plain white paper, lined into boxes for inscription, onto which would be scribbled the 'what', 'where' and 'when' of each photograph of station, shed or lineside.

From the start of my search for steam in late 1962, I was aware that the days were numbered. I never saw a 'Nelson' or an 'Arthur' in steam, nor a 'Princess'. My 'King' photographs were restricted to an SLS special and a shot of 6010 *King Charles I* withdrawn and dead outside Old Oak Common shed. (I don't know why it was there.) For almost six years I grabbed every opportunity to pursue my chosen dedication. Jane could never have realised that her modest suggestion of a weekend occupation. would develop into an obsession, for that is what it became. Journeys were plotted to include a few minutes spent where a country lane crossed a railway line; business visits were arranged to allow me to take an extended lunch-hour at an engine shed. Often Jane acquiesced in these arrangements; although sometimes, as when I kept her waiting three and a half hours in the car in Swindon, she was none too pleased! Often we planned a day's outing, with a picnic at a carefully selected site.

Slowly, I accumulated an increasing pile of scruffy pieces of paper, on which were annotated the details of my photographs. I transferred the details to the note-card provided with each Paximat magazine in which I stored the transparencies. By the time steam died, on 3 August 1968, I still had a lot of notes uncharted on scruffy scraps of paper. They remained thus for more than ten years. Not until my first book *British Steam in Cameracolour*, was contracted did I commence compiling a thorough catalogue. Thumbing through those scruffy pieces of paper reawakened my railway enthusiasm.

Once that book was under way, I began the Herculean task of creating an efficient and accurate record of my colour photographs. This is now my bible. Memory may suffice for reminiscence, but readers demand accurate detail!

To write a book for railway enthusiasts one must use extensive 'shorthand'. Initials such as GWR, LSWR, AWS, MPD, ECS (the list is interminable) – create an atmosphere of familiarity between author and reader. Yet today there exists an interest in the steam railway amongst a generation whose experience thereof is limited to such as the SVR, KWVR, NYMR and the rest! My bibliography for this book is based on the premise that the bracketed reference after the initial description will satisfy the reader who is not a dedicated enthusiast, whilst hopefully not infuriating the true steam man.

Another point of style: I cannot abide descriptions of Castles as *ex* GWR or Merchant Navies as *ex* SR. For me they remain GWR and SR engines. An A4 was an LNER locomotive, and a Black 5 merited only the descriptive LMS. We know that the railways were nationalised in 1948, but we do not need to keep reminding ourselves of the fact.

Surely, railway enthusiasm builds for its followers a range of friendships of a broader spectrum than any other occupation. Whilst recalling with pleasure my frantic photographic forays into signal boxes, motive power depots and other photogenic vantage-points, these recollections are dimmed with disappointment at my failure to record satisfactorily the men whose kindness and hospitality so enhanced the pleasure of the visit. The signalman at Pepper Hill signal box, to the west of Clifton Junction Station is a case in point. His willingness to let me take photographs from his box was but the least of his kindnesses. Working timetables were proffered and accepted, not to mention the companionship of the coffee-flask and sandwiches. He deserves a mention – yet his name is sadly lost in that inadequate storage system known as memory.

My recollection of the Ordnance Survey maps I used is fortunately buttressed by retention of the actual maps. Seventh Series One-inch Map Number 102 MANCHESTER is well-thumbed. Cloth-bound, at seven shillings and six pence it remains indispensable. What a splendid organisation – the Ordnance Survey. All that is best about Britain: superb quality, meticulous attention to detail, interest in and willingness to help – some supermarkets could learn a lesson from them! The Ordnance Survey's willingness to allow me to browse among and research into their archives, and their enthusiasm for my ambition to include some of their maps in this book, justify the not inconsiderable charges they make for servicing my requirements.

To the signalmen, shunters and shedmasters whose friendship assisted me, my thanks and my apologies where I have been unable to recall names. In some shots, people have been included for artistic reasons. I hope that somewhere some railwayman recognises himself.

The magic of the railway, for the dedicated enthusiast, lies in the unusual if not unique combination of human and mechanical interest. It is an interest about those who work on, for or with the railways, and about the contraptions they built, on which, in which, by which and with which to run the railways. Thus, for any enthusiast, the humblest suburban journey on an electric train or diesel multiple unit will bring him into contact with 'the railway' – its life, history and people. A long distance journey, even when made regularly, provides constant interest in the scene; whilst a cross-country trip on an 'unmodernised' line offers very varied fare. To make an inter-regional cross-country journey in the cab of an elderly diesel, accompanied by three different drivers, all of whom reminisce about their steam careers is, alas, an event that will soon be enjoyed only by those able to read about it, rather than experience it. My journey in the cab of an ageing Class 31 diesel from Bristol Temple Meads to Southampton, on Friday 9 May 1980, encapsulated all that Brian Haresnape meant by his phrase 'brotherhood of rail'.

Bristol is and always will be my soul city! Chapter 14 explains its general significance in railway-historical terms, and the photograph of my nephews standing beside *Misterton Hall* at Barrow Road MPD, illustrates its family link. Entering Parliament for the old Bristol North-East constituency in 1970 added another link to my Bristol 'chain': a chain which is completed by the Bristol 'Holiday Inn', a hotel built on a site selected when first I started searching for locations for hotels in the late 1960s.

So it was, at 3 o'clock on that glorious spring afternoon in 1980 that I entered Temple Meads station, that Mecca of the railway era, and headed for Platform 7, intending to catch the 15.14 Bristol to Portsmouth Harbour train, due into Southampton at 17.03. From there I would catch the 17.13 to Brockenhurst, and then the 'Lymington Flyer', due into Lymington Town at 17.50, just in time for a rendezvous with an important constituent at 18.00.

The sequence of events which followed resulted in my arriving not in Lymington at 17.50 but in Christchurch at 19.20! Over four hours trapped as a 'guest' of British Rail would have most people spitting with rage. But for me that afternoon was filled with nostalgic pleasure – initiated by the sight of an HST unit at Platform 7 as I arrived. The crowd on the platform were waiting expectantly for the HST to depart, as I walked to the point where I estimated the front of the Portsmouth train would be. Five past three: then that distinctive 'Brittol' accent passed on the glad tidings to the waiting throng – 'we are sorry that, due to the late arrival of the incoming stock, the 15.14 train to Portsmouth Harbour will be delayed.'

By this time I had reached my chosen place, and craftily sidled up to an ASLEF-badged senior railwayman, with whom I struck up idle conversation. At 15.10 the HST stock backed out of Platform 7, westwards. The driver of our train told me that we were to have the stock off the Weymouth train brought up right away 'from Malago'. My request to join him in the cab was preceded by immodest comments about my status as railway photographer, author and, *en passant*, politician. His agreement to allow me to accompany him coincided with another announcement about the delayed departure of the train.

By this time our guard had joined us. Then the platform telephone rang and as nobody answered it, I lifted the receiver and was accosted by an inspector from Exeter St Davids, who wanted a Temple Meads inspector. I told him I was only a passenger, which seemed to amuse him! . . . but I digress.

By the time our 15.14 stock had arrived from the carriage sidings at Malago Vale, in the charge of a rather tired Class 31, I had established the fact that our driver, Ron Bergelin, was a Cardiff Canton man, and that I was to accompany him only as far as Westbury. We began to talk of the steam era, and my annoyance at the train's late departure quickly evaporated. A former Canton 'Top Link' driver, Ron had fired and driven Castles and Britannias on the Red Dragon. He missed the fellowship and prestige of steam, but his feelings were by no means all regretful. 'In a freezing wind in the winter you could run from Cardiff to Paddington, cold and wet yet with hands bleeding from shifting ten tons of coal on a bad day. The prospect of the return journey wasn't so exciting, I can tell you'.

We talked of the status of the railwayman: yet as we gathered speed through Sydney Gardens, Bath, people waved to us. 'Last month I was offered more money to work in the central market in Cardiff than I get after 38 years on the railway,' he said, 'but, I couldn't just change from this, could I?'

How undervalued is the loyalty and dedication of the BR footplateman, weaned before the grouping and apprenticed on the 'Big Four'.

As we left the GWR main line at Bathampton, turning south towards Bradford-on-Avon, the sight of the Avon Valley from the cab was glorious; scenery for which men fought and died.

'You a Labour or a Tory man?' Ron asked. I told him and he told me he was a constituent of George Thomas, Mr Speaker, which seemed fitting. I admire George as a fine man – in good company with GWR engine drivers, men whom I have always held in the highest esteem.

The line twists and turns with the river, and we proceeded gently through Freshford, Avoncliff and Bradford, over Bradford West Junction where I noted the rusty state of the line towards Melksham and Thingley Junction, now freight only, until we eased to a halt briefly at Trowbridge. The GWR lives on hereabouts, and as we clattered into Westbury the eye was gladdened by numerous GWR lower-quadrant signals. As Ron and I took our leave my pleasure was tinged with sadness at the realisation that such company was becoming scarce.

As the new driver exchanged a brief word with Ron Bergelin on the platform at Westbury, I hovered by the cab door, hoping for permission to stay. The new driver had a fine head of red hair and beard to match. He assured me, in a quiet but friendly voice, that I was welcome. Even as he settled into his left-hand driver's seat, my eye alighted on another reminder of the greatness that was the GWR: alongside us was Class 47 No. 47079 *G J Churchward* – arguably the father of the modern steam engine.

From Westbury we took the left-hand road, passed another Class 47, crossed the bridge over the Berks & Hants line at the Westbury cut-off, gathered speed through Dilton Marsh and began the climb to Warminster. Driver Michael Richardson, knew there was little point in trying to flog his steed – there was nothing there! Pausing briefly at the neat station at Warminster, we began to talk as we became acquainted. The track was rough here, very rough in places.

'It's only relaid when there's been a derailment', said the driver.

By the evidence of my eyes, this was not an infrequent occurrence.

I asked him how long had he been at Westbury?

'Seven years.'

'And where were you before that?'

'Bromsgrove.'

'On Lickey Bankers?'

'Yes.'

'What did you think of the 9Fs?'

'Excellent engines,' he said. 'We had Great Western panniers, and before that we had the Jinties.'

I asked him if he had ever seen 'Big Bertha', the unique 0–10–0 built by the Midland Railway for duty on the Lickey Incline in 1919, a class of one, withdrawn by BR in 1956.

Not only had he seen her, he had been on the footplate. You need to be a railway enthusiast to appreciate my feelings at being in the cab with a man who had been aboard 'Big Bertha', as 58100 (or 22290, to give her her LMS number) was known.

Down the grade we gathered speed, passing the site of Heytesbury and Lodford stations. The Wiltshire Downs looked glorious in the warm sunshine, the cattle well-fed and with ample space. At Wylie we passed the trim signal-box, still open, and soon we were caught in the lens of a railway photographer, whose cheerful wave we returned.

By now I was noticing the oddity of two sets of milepost. The older cast-iron ones had been repainted in dark yellow and black. It was the SR black-on-white ones which reminded me that we were approaching another famous railway location, Salisbury. As we reached Wilton we crossed from GWR to LSWR tracks about a mile west of the former junction. Suddenly the magnificence of Salisbury Cathedral came into view, totally uncluttered by the roadside paraphernalia that comes between the motorist and that magnificent building.

Recalling Merchant Navies accelerating westwards after their Salisbury engine-change during the days of steam, I realised I would soon be saying farewell to Michael Richardson. As I pondered on the character of his replacement, the change from GWR to SR was self-evident in the latticed signal posts. We saw two youthful spotters sitting astride a lineside gate, and then we were on the final approach to Salisbury station. Opposite the West signal-box I noticed a snow-plough affixed to a six-wheeled steam engine's tender.

As we came to a halt, I bade farewell to another man who had become a firm friend in a short space of time. On the platform he exchanged words with driver number three, Reg Weeks, a Fratton man who bade me welcome on his footplate – I mean cab. This craggy man reminded me how late we were, by telling me that a girl on Salisbury station had regaled him with her problem – a night-club engagement in Brighton that evening and the worry of a lost connection at Fratton!

We emerged from the tunnel at Tunnel Junction and Reg began to chat about speed limits. Like all enthusiasts, our minds linked Salisbury inextricably with the ghastly accident on the night of 30 June 1906. Reg recalled it to mind by commenting that he felt the 10 mph speed limit was a little over-cautious; yet what historian can erase the memory of that summer night long ago, when for some unknown reason Driver Robins hurtled through Salisbury with the American Liner Special from Plymouth, and the train was completely wrecked. 24 of the 43 passengers were killed. That tragedy will never fade from Britain's railway folklore.

At Tunnel Junction we turned off the LSWR West-of-England main line and again headed south-east. Reg and I searched for, and found, the precise location of the former Alderbury Junction, all but obliterated by the realignment of the A36 road. (Never more will trains run to Downton, Breamore, Fordingbridge, Daggons Road and Verwood.) Before Dean station we saw some charcoal-burners in a clearing in a wood beside the line. Their ancient craft seemed in keeping with this interesting cross-country line. Immediately thereafter we slowed to 25 mph for a mile, the result of another derailment some six months before, when a laden hopper had become derailed, without severing the vacuum-brakes.The hopper's wheel had smashed against the concrete sleepers for a mile before the driver became aware of the problem. Perhaps the damage to old-fashioned wooden sleepers would have been less destabilising. . . .

We passed through Dunbridge. Reg pointed out such items of interest as the automatic level-crossing barrier activators. As we passed Romsey and the delightful little signal-box at the junction with the Chandlers Ford line to Eastleigh, I knew that my reverie must soon end. As Reg pointed to the site of the former Nursling station and commented on the brilliance of the yellow mustard in large quantity hereabouts, the link between past and present brought a slight lump to my throat. Soon the sparkling water of the River Test was beside us. Then we were under the bridge and joining the electrified tracks at Redbridge, as a Southampton to Bournemouth 'stopper' pulled away from Redbridge station. With very real regret I said goodbye to Reg at Southampton and shook his hand before jumping down from the cab.

That night I browsed through *Steam Around Bristol* by Rex Coffin, the first picture in which is a shot of No. 4080 *Powderham Castle* departing from Platform 7 at Temple Meads, the very platform from which I began the journey detailed in these last paragraphs. Also, from that very spot I photographed the departure of the Bristol-Portsmouth train shown in plate 60.

Four days after my journey from Bristol to Southampton, I enjoyed another footplate journey, from Portsmouth to Waterloo. Again I had the pleasure and privilege of listening to a veteran railwayman recounting tales of yesteryear. What men they are! This time my companion wanted to remain anonymous, although he told me of his 45 years railway service, spent mainly at Guildford steam shed, and latterly at Farnham electric depot. He had followed the traditional career pattern from cleaner to fireman, ultimately, to driver.

Life is about people, and the steam railway was never short of characters. That Tuesday, 12 May 1980 was a hot and gloriously sunny day, yet my driver on the 'Portsmouth Direct' would not have dreamt of driving his train in his shirtsleeves. He was sired by the railway, his father being an LSWR man who revered Dugald Drummond, Chief Mechanical Engineer (CME) of the South Western from 1895 to 1912. Drummond, so the driver told me, was respected in life and revered in death by some but not all of the SW drivers. If he observed an engine blowing-off under the roof at Waterloo he would issue a stern reprimand both for the waste and the noise. The drivers, wary of Drummond's sorties, would run their engines from Nine Elms shed to Waterloo with only two or three inches of water in the boiler in an effort to prevent this 'misdemeanour.'

Drummond was buried in Brookwood cemetery, alongside the LSWR main line, and for years afterwards some drivers would doff their caps to him as they sped by. Yet many SW men feared and hated Drummond; some reputedly put half a ton of brake-blocks on his grave to ensure he stayed there!

As we headed towards London, my host commented on points of interest. At Liss he pointed out the huge concrete buffer-stops at the terminus of the Longmoor Military Railway (LMR), apparently the ultimate means of stopping their numerous inexperienced drivers! 'Six months training and they thought they were engine drivers', he said. 'It's like the chaps today, they aren't railwaymen in my sense of the word!' Here was a man who unashamedly felt frustrated by the demise of steam.

'There is no relation between your knowledge, experience and effort, and the standard of the job you've performed,' was the way he described the driving of an e.m.u.

Himself a fan of Drummond's engines, particularly the T9s, he also had good words for Maunsell's U moguls, but not much favourable to say about Bulleid's Q1s. 'Ugly outside – and uncomfortable to drive!'

As we pulled out of Haslemere and descended Haslemere bank he relived the task of bringing seventy loose-coupled wagons up the bank, and then reminisced about the importance of the partnership between driver and guard in dealing with heavy loose-coupled freight trains on some of the switchback routes in Surrey and Hampshire.

As we approached Guildford he suggested that discretion demanded I pass back into the train, lest an inspector appear. At the Woking stop however, as I looked out of the window of my compartment, he invited me forward again to introduce me to another driver.

Again I listened to the resonant tones of another long-serving railwayman, this time a Fratton man. The two of them had served in the army together in India, and had some incredible tales to tell! He was a Devonian, having started his railway career at Plymouth Friary before the war.

Researching this book has been more than enjoyable; through it I have met one of the key men of the steam era, alive, well and living in Lymington. Kenneth John Cook – KJC to those in the railway world – was apprenticed to the great G J Churchward when he joined the GWR at Swindon in 1912. Yet in 1980, sitting by his drawing-room fire, his matter of fact recalling of his first detailed discussion with Churchward – 'in Stanier's office' – left me feeling like a privileged party to the living past. Notwithstanding affable conversation with past and present prime ministers, kings and princes, television personalities or other eminent folk, his company and presence will stand the test of time among my memories.

KJC was with Churchward minutes after Churchward's subsequently fatal accident in the swirling mist at Swindon. 'I had to tell his household of the tragedy. He was a bachelor, his housekeeper, maid and the three retired railwaymen who did his garden were shocked.'

His personal recollections of Collett, under whom he worked as Locomotive Works Manager at Swindon, must be recorded for posterity – a task we shall try, together, to fulfil ere long.

As I have gone 'In Search of Steam', it has become clear that there lies untapped in Britain yet, an untold treasure trove of memories and recollections among the men of the steam era.

Thank you KJC – we shall meet again. Thank you Brian Haresnape, for advice. Thank you Derek Cross – not only for bothering to read my text, but for being honest with your comments, some of them probably libellous but none the less tremendous fun! The *Cross Critique* would have made an excellent Chapter 16! Derek is not given to flattery – at least not by intention. Therefore my ambition is no more than that you may be able to read this book and say as he did . . . 'Going through your book my own experiences have come back vividly to my mind'. I could not hope to achieve more . . .

60110
60117
60117

1 End of Steam Around London

Railways were more than steam engines. Whereas roads and airports grow to serve the population centres they adjoin, the railway itself created many towns and cities. Early railway development had a profound effect on urban architecture. Stations were seen and used by millions. Yet to those who study the world of railways, there remains a wealth of architectural interest in the buildings provided to service the steam engine. Many engine sheds with which enthusiasts had become familiar, died when steam vanished. Some were smashed into the ground within weeks. Water towers, coaling plant, turntables, track, ashpits, shed buildings and offices – all were reduced to rubble, or burnt. My sorrow at their passing refuses to allow me to apologise for lamenting their end, or for recording the fate of a few of my favourite locations.

In searching to recapture the mood and atmosphere of a shed with which I was once familiar, I revisited Feltham on Wednesday, 2 January 1980. It was a macabre journey: as I roamed the huge, derelict and desolate site, more wilderness than man-made, I was lost in my memories; but the passing of an electric train on the Waterloo/Staines/Reading line awakened me from my reverie.

A Class 37 Diesel hove into view, hauling bogie oil-tanks; and I remembered Feltham's past role as one of London's main freight concentration centres. Those cross-London freights always interested me, and that passing Class 37, which I fumblingly sought to photograph with my new Cannon 100/200 mm lens, set me to write this chapter about the end of steam around London.

Before plunging into any detail, it is necessary to remind ourselves of the attitude of British Rail to steam in the 1960s. Unfortunately it was akin to the relationship between a Maharajah and an Untouchable. The last thing the Maharajah wanted was Untouchables in his capital city – so steam was exiled

PLATE 1. The original GNR 'engine stables' in London were built in 1850. A large part of this original building remained in use for engine repairs, until the depot, known to those associated with railways as 'Top Shed', closed in 1963. From their construction in 1922 until their demise, the Gresley Pacifics dominated and indeed epitomised Top Shed, seen here in March 1963. The shed closed, and an era ended, three months later.

from London at the earliest opportunity. It remains one of the great ironies of railway history that the last main-line steam service from London ran on the Southern – the Railway which, under the leadership of Sir Herbert Walker, had sought to eliminate steam in the 1930s. In those days, whilst GWR, LMS and LNER based their motive power on steam, the Southern Railway electrified vigorously. Only Bulleid and Hitler ensured that steam survived and prospered on the SR. The three other railways often sneered at the Southern – 'merely an extended tramway' was one comment.

Ironic indeed that main-line steam – or any steam at all – survived at Waterloo to outlast every other London terminus. My own record of photographic visits to the main motive power depots serving the London termini, tell their own story. These dates record only my own last 'photo-call', not the actual last day of steam, but they paint the all-too-familiar picture of steam's elimination, as one by one BR's regional managers sought to remove the motive power which they, quite unfairly, tried to blame for the railway's financial misfortunes.

Top Shed (Kings Cross)	LNER	March 1963
Camden MPD (Euston)	LMS	28 April 1963
Stewart's Lane MPD (Charing Cross/Victoria)	SR	13 July 1963
Watford MPD	LMS	14 May 1964
Willesden MPD	LMS	5 July 1964
Old Oak Common MPD (Paddington)	GWR	23 January 1965
Southall MPD	GWR	6 June 1965
Feltham MPD	SR	6 June 1965
Cricklewood MPD	LMS	30 November 1965
Nine Elms MPD (Waterloo)	SR	9 July 1967 (London's last day of steam)

When I started my search for steam to photograph, at the end of 1962, I could hardly have been less well equipped. Not since my schooldays had I exercised my deep interest in railways, and my knowledge was rusty. It was during a visit to my wife's parents at Burnham-on-Sea, in S & D country, that I made the first conscious decision to *look* at the railway. The short journey from Burnham to Highbridge reawakened my long-dormant enthusiasm and once rekindled, the chase was on. By the time we returned to London from Somerset that weekend, my mission was clear – find the engines.

Where do you begin? Sitting in our flat in Chiswick, it was as though I had made an abstract decision to become a mountaineer – without knowing where the mountains were, or how to find them. Running through my head were memories of my childhood source of information, the *Wonder Book of Trains* – ridiculous thought! Then I remembered *Railway Magazine*, plus Ian Allan's unique books of numbers. I recalled ticking-off the numbers of Southern PULs and PANs on visits to the railway line during school walks from my prep school, Falconbury near Bexhill. (We never seemed able to find 6-PUL No. 3014, ENID!)

That was it! First stop was a bookshop. The next requirement was a map; then the purchase of *Railway Magazine*, and reference to the section entitled 'Locomotive Notes'. Then I joined the Railway Correspondence & Travel Society (RCTS), to which I still belong. Future railway historians will have cause to thank the RCTS and its members for their diligent enthusiasm and enthusiastic diligence!

The chronological decline of steam around London began with the LNER. Kings Cross was an obvious early destination. Joining the 'platform ticket brigade' I found myself alongside youths of tender years, number-books in hand, who reminded me of my early days standing at the end of Platform 2 at Brighton. The few remaining LNER Pacifics interspersed amongst the Deltics and Type 4s were like sweet sultanas in a very doughy pudding. Ten minutes on the end of that platform at Kings Cross in November 1962 made me determine to seek access to the packet of sultanas! Top Shed Kings Cross became my target.

It is sometimes said that life is not about *what you know* but about *who you know*. This presumes that who you know is a matter of divine ordinance, which is false; it is a matter of your determination. There were, and still are, three ways of gaining access to an engine shed. You can obtain an official permit from British Rail: you can arrive 'on site' and seek permission to look round from the shedmaster or duty foreman; or you can trespass. Rarely if ever was the latter necessary, thanks to the generous assistance of BR staff at all levels. Never having found it difficult to get to know people, I found little problem in establishing contact with those in a position to issue shed passes, or lineside permits.

A study of 'Locomotive Notes' and the RCTS monthly magazine *Railway Observer* clearly indicated the impending demise of LNER steam in the London area. Furthermore, the LNER was the railway with which I had had least contact, thus increasing its appeal and mystique.

Although in March 1963 steam was still seen on occasional inner and outer suburban services from Kings Cross, the day of the diesel had well and truly arrived. Likewise, the main-line services still had some steam haulage, but clearly steam was on the wane when I made my pilgrimage to Top Shed, Kings Cross – the Mecca of LNER steam in London.

I suppose Gresley's A4 Pacifics are more instantly recognisable than any other steam engine ever built. Mallard's world steam speed record was a major publicity coup for the LNER. Thus the sight of 60007 *Sir Nigel Gresley*, albeit undergoing light repair, inevitably stimulated my sense of occasion. Als, A3s, V2s, B1s and even a K3/2 were scribbled onto my notepad, plus an engine that became, all too briefly, a particular chum; the A2/3 Pacific No. 60523 *Sun Castle*. One or two WD 8F 2–8–0s and BR/STD 9F 2–10–0s added a more workday image to the scene. Top Shed's renowned cleaning squad were clearly still on parade.

That was my first and last visit to Top Shed. With the advent of the 1963 summer timetable, steam at Kings Cross was eliminated. The last weekend of steam activity saw a hectic exodus of Top Shed's stock, and on the final Friday evening, 14 June, ten out of sixteen northbound main-line freight and passenger trains were steam-hauled. On the following day, Kings Cross' two A4s, 60017/25, worked the 09.05 to Newcastle (Tyne Commission Quay) and the 09.20 *White Rose*.

One by one the remaining engines left, some 'light engine'. Final evidence of the end of one of the world's most famous engine sheds came as A3 60112 *St Simon* from Grantham (34F) towed Kings Cross' stationary boiler, K3 No. 61912, north.

The last scheduled steam working from Kings Cross was the 22.45 to Leeds on 16 June 1963, by Doncaster A1 No. 60158 *Aberdonian*. *Sun Castle* was withdrawn that month.

Tabulating the exact decline and fall of the London-area sheds listed on page 22 would fill more space than I can spare, so my fleeting impressions must suffice. Cricklewood was Jubilee's, plus an ancient Johnson Midland 0–6–0T passing the shed, hastily photographed hauling new cars which now look antique themselves. The engine was one of those fitted with condensing apparatus for working in the London area.

Camden was 'Coronation' Pacifics inside the shed. Stewart's Lane, across the river, was on its last legs in July 1963, with only an N mogul in steam, although I photographed a (dead) Schools 4–4–0 No. 30928 *Stowe*, now undergoing restoration on the East Somerset Railway at Cranmore, plus another relic of yesteryear, one of Wainwright's South Eastern and Chatham H class 0–4–4T engines.

Stewart's Lane once had the highest allocation of steam stock on the Southern Railway. Following rationalisation by the SR in the early 1930s, the shed was selected to house the main steam engines to service the former LC&D, SE&C and LB&SC lines which were not being electrified. Given a new roof in 1934, the depot had an allocation of some one hundred and seventy engines.

By July 1963, the Lord Nelsons, King Arthurs and even the Bulleid Pacifics were merely ghosts of Stewart's Lane's past.

Willesden was both cramped and crowded, and even as late as July 1964 rewarded one with named engines, which had become rare through the rapidity of withdrawal of once-familiar locomotives. There were Britannia, Jubilee, Coronation and Royal Scot 'namers' still in evidence, as well as Fowler 4F 0–6–0s. Willesden Junction was one of those railway locations that seemed as permanent as time itself – even our school special stopped there – but today the electric trains from Broad Street to Richmond, passing above the West Coast electrified main line from Euston at Willesden, now provide more memories of times past than does the main line itself.

Churchward's Old Oak Common was incomparable. Its palatial roundhouses contained four 65 ft diameter turntables, all interconnected by a line of rails so that engines could move from one to another and thus gain access to any corner of the complex. Opened in March 1906, Old Oak Common was approaching its eclipse of steam in January 1965; this was arranged for 22 March, when the by then reduced steam stock was transferred to Southall. Demolition of the main building had started a year earlier, and some macabre pictures presented themselves to my camera as the demolition work proceeded.

Southall shed was much less grand than Old Oak Common. On 6 June 1965 I paid my last respects to a great engine, 5042 *Winchester Castle*. It was the last GWR-built 'Castle' to remain in service, being withdrawn a few days afterwards. Never shall I forget that sight: it was shameful. One of the Southall drivers kindly photographed me on the buffer-beam that day. Southall lost its official steam allocation two months later, and was closed to steam at the end of the year – the year that heralded steam's elimination from the GWR.

My final roll-call of the London sheds brings me back to Feltham, and to Nine Elms. Both were former LSWR depots, although Feltham was not opened until just after the 1923 grouping. Feltham was mainly a freight depot, from which motive power was supplied for goods trains all over the ex-LSW main lines, as well as for the cross-London transfer freights for which Urie designed the powerful and unusual G16 4–8–0T and H16 4–6–2T engines, which were based there.

The depot was part of a large freight complex, including the 'hump' marshalling yard. Railway activity covered an impressive

PLATE 2. Plodding along the 'Premier Line' from London, unkempt Stanier 8F 2–8–0 No. 48754 trundles heavy freight north near the site of the Great Train Robbery, in pre-electrification days at Tring cutting, on 23 June 1963. *(Above.)*

area. In addition to the shed, coaling plant, turntable and familiar paraphernalia, there was the landmark of Feltham's clock-tower in the marshalling yard.

Returning to the site today is, as I have intimated, a weird experience. The shed closed with the end of steam in 1967; the S15s have all gone. The depot has long since been demolished and the track lifted. Scattered about this huge acreage of derelict land are lumps of concrete; telegraph poles and lengths of rubber hose lie trapped in the undergrowth. You kick at a rail chair-block and it clatters into a rusty tin lamp; ash pits are filled with rubble, a stool stands alone, perhaps left over from the canteen of this once active place.

Standing cold, gaunt and forbidding against the winter sky is the shell of the clock-tower. Clambering over debris, I reached it and climbed the stairs, wondering what I might find. The wind whistled through the empty building – was that a hiss of steam, a clank of a piston-rod?

Whilst Feltham was predominantly devoted to freight, Nine Elms was the principal passenger locomotive depot of the LSWR. Whereas Feltham was a modern shed on a new site, there had been an engine shed at Nine Elms since the dawn of the railway era. By 1910, when the 'new' motive power depot was

PLATE 3. Now derelict wasteland, Nine Elms North Yard was its busy self on 3 July 1964 when Maunsell 'Q' 0–6–0 No. 30545 was shunting. This shot was taken from a passing electric train running on the Windsor lines into Waterloo. This engine was the last survivor of a class that was never very popular; she was withdrawn in May 1965. (*Above*).

constructed, Nine Elms had grown into a vast 25-road shed on the south side of the main line, whilst to the north, reaching to the river, was the acreage of a huge goods yard.

With electrification came reduction in numbers of engines, although the LSW main lines to Southampton, Bournemouth and Weymouth, and Salisbury, with attendant empty-stock workings between Waterloo and Clapham, remained largely steam-hauled until 1967, by which date Nine Elms was the last remaining steam shed in London.

9 July 1967 saw steam's final day of scheduled service. It went out with a whimper rather than a bang in the capital of the nation that invented the steam engine. As the engines came onto Nine Elms shed for the last time, men who had spent a lifetime with steam, bemoaned the decline of the pride, glory and enthusiasm, the very life itself that the steam engine epitomised. Dank depression was everywhere; a grim, melancholy mood gripped those men whose lives would never be the same. Only their memories would remain tomorrow. Much of the site is now let to the Covent Garden Market Authority and forms the flower market – another dismal example of the emasculation of tradition, and its replacement by faceless, featureless concrete.

Memories of steam in and around London are not restricted to

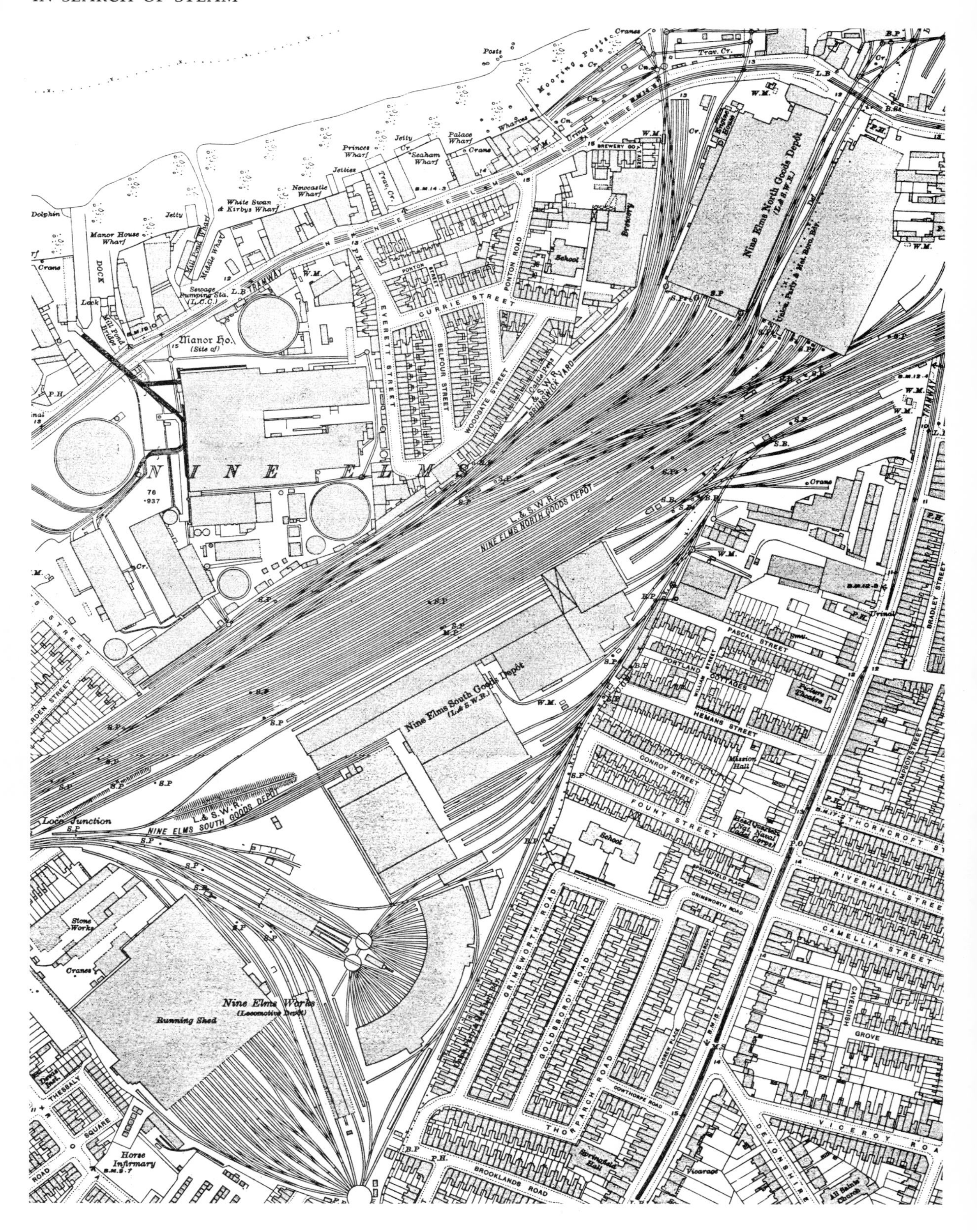

Nine Elms Works
(Locomotive Depôt)
Running Shed
Nine Elms South Goods Depôt
(L.& S.W.R.)
Nine Elms North Goods Depôt
(L.& S.W.R.)
L.& S.W.R.
NINE ELMS NORTH GOODS DEPOT
L.& S.W.R.
NINE ELMS SOUTH GOODS DEPOT
NINE ELMS
Loco Junction
Manor Ho.
(Site of)
Brewery
School
Stone Works
Horse Infirmary
Picture Theatre
Mission Hall
Vicarage
All Saints' Church
EVERETT STREET
BELFOUR STREET
CURRIE STREET
PONTON ROAD
WOODGATE STREET
PASCAL STREET
PORTLAND COTTAGES
HEMANS STREET
CONROY STREET
FOUNT STREET
GRIMSWORTH ROAD
GOLDSBORO' ROAD
THORPARCH ROAD
BROOKLANDS ROAD
THORNCROFT
RIVERHALL STREET
CAMELLIA STREET
VICEROY ROAD
DEVONSHIRE
BRADLEY STREET
THESSALY SQUARE
TRAMWAY
Manor House Wharf
White Swan & Kirbys Wharf
Newcastle Wharf
Princes Wharf
Seaham Wharf
Palace Wharf
Dock

Fig. 2 (page 28) and Fig. 3 (page 29).
The actual changes in the landscape are self-evident in these maps of Nine Elms: yet the changes in style and terminology are as informative as the content. The map dated 1916 denotes Nine Elms Works as the 'Locomotive Depot'. By 1952 the Works had disappeared – transferred to Eastleigh in 1909. Again, the early map indicates the motive power depot as the 'Running Shed', referred to as 'Loco Shed' on the later map. Rail transport at its zenith coincided with horse-drawn vehicles on the streets: thus the 'Horse Infirmary'. Alas, to-day, trams run no more along the Wandsworth Road or Nine Elms Lane. Features familiar to steam enthusiasts, such as turntable and coal-hopper, have gone, too, on the latest map. Social historians can discern from these maps the impact of the blitz upon the area. They may also ponder whether Tidemore House sees more, or less, vandalism than the dwellings on the site in yesteryear.

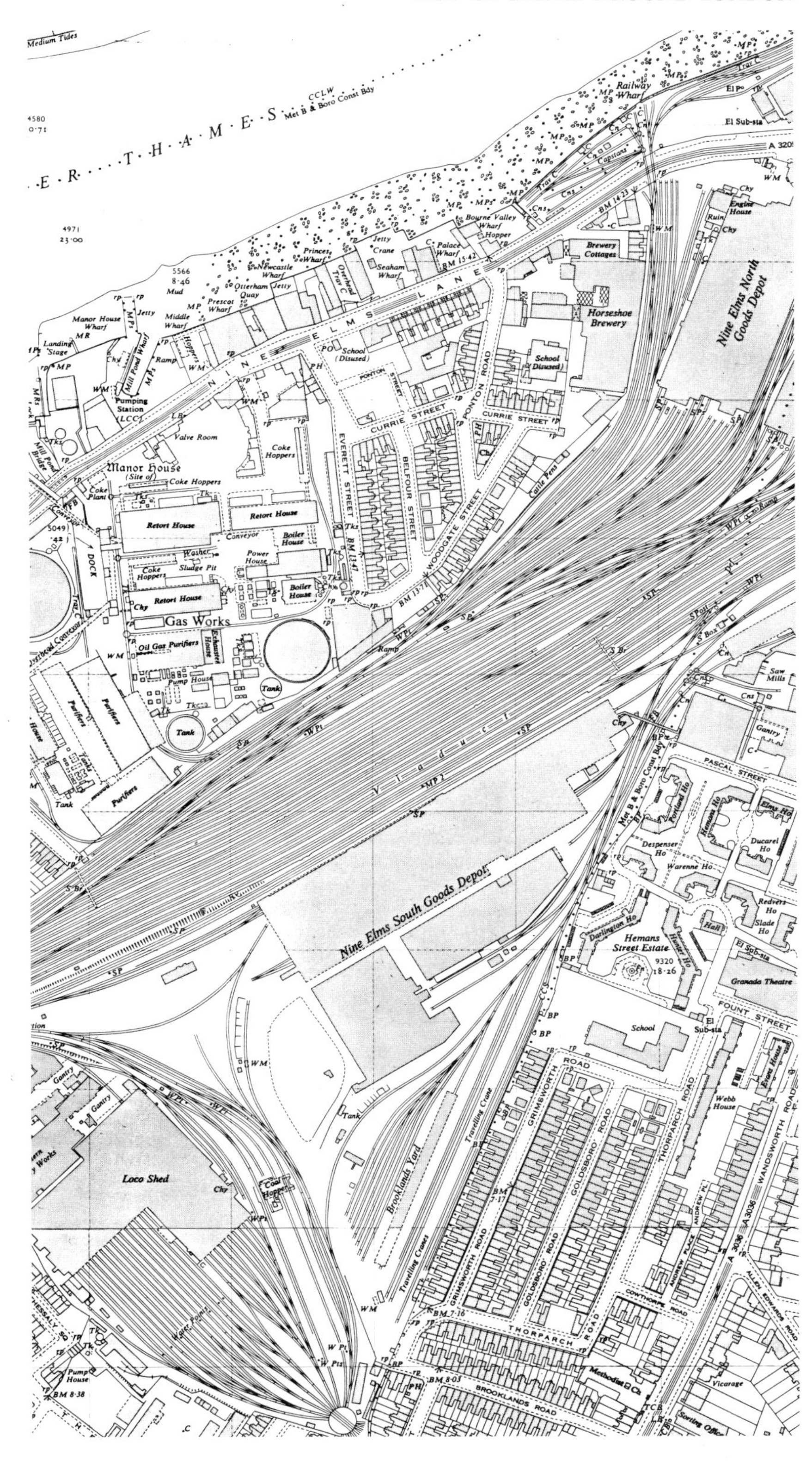

PLATE 4. Saturday morning; drifting down the grade towards Egham in January 1964, Maunsell S15 4–6–0 No. 30844 heads a freight from the Western Region via Reading. She is bound for Feltham Yard. (*Following pages.*)

Fig. IV. Tring Cutting, 17 June 1837.

the engine sheds. There are the great stations, cuttings and tunnels; and the suburban scenes through suburbs whose growth reflects the urban transformation wrought by the railway. Out beyond the suburbs, on the approaches to the city, great works were conceived by our Victorian forebears to bring the railway to the capital. Men in their thousands struggled, and some died, to build giant cuttings like Tring and Sonning. Bridges and viaducts have added to, rather than detracted from, the urban landscape. The graceful arch of Brunel's bridge over the Thames at Maidenhead or the viaduct over the River Colne near Watford have outlasted even the steam engine. To go in search of the railway around London can still reward the eye.

Although the great London termini gave steam proximity to the daily life of the city, its visibility was generally restricted to those on railway business. Crossing the river by Waterloo, Hungerford or Chelsea bridges, one saw the tell-tale evidence of steam, and yet there was one place in the capital where steam's presence was brought into intimate contact with motorists in an altogether surprising and, to the unwary, disconcerting way. It was at the point where London's 'secret railway', the West London Extension Railway (WLER) meets the A4 Great West Road, near Earls Court.

PLATE 5. Steam in the Royal Borough of Kensington, at Olympia on the former West London Extension Railway. Fairburn 2–6–4T No. 42118 passes the Exhibition Hall on a transfer freight from Willesden to Feltham in November 1962.

The WLER has a fascinating history. In 1836 a railway was authorised with the grandiose title of the Birmingham, Bristol and Thames Junction Railway. It failed to reach any of these three points, but was a three-mile branch line from the Birmingham Railway at Willesden to the terminal basin of the Kensington Canal. In 1837 the diversion to Paddington was authorised and in 1840 it changed its name, more appropriately, to the West London. Steam survived on the WLER until 1971, operated not by BR but by London Transport. In addition to the cross-London freight trains that still form the backbone of the line's activity, one of London Transport's former GWR pannier tanks could occasionally be observed. Serviced at Lillie Bridge depot, it could be seen simmering at the signals where the line passed under the Cromwell Road Extension. At this point, motorists sitting in traffic jams could suddenly be confronted by billowing white smoke from the WLER tracks below. The incongruity of the sight of a steam engine at Olympia may be judged by the shot of a Fairburn 2–6–4T passing the Exhibition Hall on a crisp November day in 1962. Until quite recently the sight of GWR semaphore signals beside the six-lane motorway link to Shepherds Bush was a welcome reminder of the WLER, whose tracks saw the last steam around London.

2 Lancashire

'What's in a name?' To most of us Lancashire means cricket or cotton or the Wars of the Roses. Burnley and Preston mean football or maybe motorways. North East Lancashire means industrial deprivation or the jargon of some infernal socio-economic planning department. But Rose Grove and Lostock Hall mean much more, to me.

Once the 1955 British Rail modernisation plan had been accepted, its implementation meant that British Rail steam was doomed. At the time, the end of the era seemed a distant horizon – rather like those insurance company advertisements that show a fresh-faced youth who, having got a job, soliloquises smilingly 'no pension – never mind'. The advertisement then shows him ten years later, applying for another job and saying 'no pension – oh well'. Ten years later he thinks, 'no pension – perhaps I should do something'. Finally the advertisement shows a distraught and haggard man throwing up his arms in despair and saying 'no pension – what *am* I going to do?' – the *am* being heavily underlined.

So it was with steam railway enthusiasts. As the 1950s slipped into the 1960s our minds were forced to consider the unthinkable – a railway without steam. Not until 1962 did this harsh fact impinge itself sufficiently on my mind to generate action. In the ten years between leaving school and contemplating this melancholy fact, my interest in railways had lain nearly dormant.

As I have said, it was Jane who rekindled my interest in steam. Married in 1961, girls, beer and rugger had perforce become spectator sports, and by autumn 1962, after a year of marriage, my sallow complexion prompted Jane to propel me 'into the fresh air'. Thus was my enthusiasm for steam reincarnated. With a camera forever by my side, I spent the next six years chasing around Britain searching for the constantly diminishing areas of steam operation.

Let me return to 1955 and Dr Beeching. At first the demise of

PLATE 6. Rakes of empty wagons were a common sight in industrial Lancashire, frequently hauled on short trips by LMS 3F 0–6–0Ts such as No. 47284, seen heading past Royton Junction, north of Oldham, on 22 April 1964. The bleak landscape was somehow made less dank by the friendly, jaunty sound and sight of engines like the Jinty.

PLATE 7. 'Steam Lowery'. Industrial Lancashire characterised by mill-chimneys, viaduct and Stanier 8F. Alas, modern technology has replaced the character, the atmosphere and many of the jobs with sterile, capital-intensive machinery.

steam not only *seemed* distant, it also started slowly. In 1959 British Rail were still building steam locomotives and almost five hundred steam sheds were still operational. However, odd vehicles called 'diesel multiple units' soon made their appearance on the Western Region. By the time BR built their last steam engine, *Evening Star*, at Swindon in 1960, BOAC (as it was then) had introduced their first Boeing 707, and steam still seemed king – except to those closely observing the changing order.

From 1960 onwards, the trickle of the mid to late 1950s turned into full flood as class after class of once famous engines was exterminated. By the time my rekindled enthusiasm caused me to join the RCTS the phrase 'class extinct' had tolled the death of many a famous, loved or respected class of locomotive.

If my dream of 'steam in my lens' was to become reality, I had to plan my life unashamedly around the remaining strongholds of steam. To do this whilst retaining a contented spouse and a satisfied boss, necessitated two conditions – a patient and understanding wife, and job mobility. Good fortune supplied both. I have always shared the view of the late Bishop Treacy that railway enthusiasts need understanding wives. On the employment front, as Sales Director of the May Fair Hotel, my task was to keep the hotel's bedrooms filled; and, after all, businessmen from areas of Britain with steam-trains operated paid their bills as promptly as those from areas that had fallen prey to the dastardly diesel!

And so, to Lancashire! As branch lines closed around the country, and as more and more diesels emerged from the manufacturers, the new men at BR vied for the title 'Champion Steam Eliminator'!

It was grim to hear the successors of Dean and Churchward gloat that steam was eliminated from the Western Region, in 1965. What faceless creatures gloated thus! As 1968 dawned, it was Lancashire, heartland of the industrial revolution, that was clearly going to claim the noble title – 'Last Stronghold of Steam'.

Arguably the last steam-hauled commuter services were from Manchester Victoria, via the erstwhile Lancashire and Yorkshire, and London and North Western lines absorbed into the LMS in 1923. In the spring of 1964 one could still be fortunate enough to witness a steady stream of steam-hauled trains plying reliably between Victoria and Preston, Blackburn, Wigan, Horwich and Bolton – on lines that today boast at best decrepit d.m.u.s – or more likely merely weeds and echoes of the past; echoes of steam.

Whatever the claims of Birmingham, I consider Manchester to be England's second city. The city has a proud history and its people a ready wit, and it is at the heart of a region where Railway Developments came in capital letters. The great adjoining stations of Exchange and Victoria have a turbulent history, reflecting the rivalries between the early companies, such as resulted in the Battle of Clifton Junction (see Chapter 8). They shared the longest single platform at any railway station in Britain. To the east of Victoria station is the fierce slope of Miles Platting Bank. To assist trains up the incline, banking engines were stationed alongside a wall that was blackened with the grime and smoke of generations of engines.

The banking duties in the dying weeks of steam were in the hands of Stanier Black 5s. I shall always remember the sight of simmering Black 5s at Manchester Victoria. Nor shall I forget the sight of a Caprotti Standard 5MT 4–6–0 blasting furiously through the station prior to charging the bank. What a sight – what a sound!

Unfortunately, in August 1967 my camera was suffering the after-effects of a mauling by our Labrador Sam – something I did not discover until two reels of film were ruined.

Never believe the tales about rain in Manchester; I was often favoured with the sun on my visits to Lancashire.

The destruction of the steam stock, and the final elimination of steam haulage was like a macabre game of inverted chess. The

PLATE 8. Low maintenance costs, reliability and widespread route availability ensured that Sir William Stanier's 'Black 5' 4–6–0, introduced by the LMS in 1934 ushered out the era of steam in Lancashire, indeed in Britain. At Newton Heath, Manchester, on 3 December 1964, No. 44689 restarts a freight from a signal check. This was one of the final engines, built at Horwich, in 1950, after nationalisation and has Timken roller-bearings on the driving axle. *(Following pages.)*

44689

specialist pieces – King Arthurs, Kings, A4s and Coronations – went first. The more numerous units, the pawns, remained until the end. Yet, keeping with my chess analogy, can it have been kindly fate that decreed that 'King Stanier' should be the last man on the board?

As the end approached, it became clear that the final echoes would sound, not from the chimney of the post-war BR standard designs, but from the 8Fs and the Black 5s of the 1934–1935 era.

As the Manchester sheds like Patricroft and Newton Heath closed, followed by Bolton in 1968, we were left with just three from which to foreclose the glorious, never-to-be-repeated era of steam: Carnforth, Lostock Hall near Preston, and Rose Grove, Burnley. For me, visits to remote Carnforth were hard to justify on the basis of visiting prospective guests for the May Fair Hotel. However, both Preston and Burnley were near enough to centres of commerce and industry to justify my spending time there. Thus, as that fateful day of 3 August 1968 approached, after which regular scheduled steam would be but a memory, my need to visit Burnley and Preston became more pressing! More and more businessmen from the area received my rapt attention. . . .

PLATE 9. Another 'Black 5' hustles a freight, tender first, past Newton Heath Carriage Works, her clean lines evident even from this angle. Newton Heath was well-endowed with railway installations – nearly all now but a memory. In contradistinction to the photograph in plate 8 this engine, No. 45017, was one of the original Crewe-built engines, entering service in 1935. *(Above.)*

PLATE 10. The BR Standard Class 5 4–6–0 was closely modelled on the Stanier classic, with few if any more effective or efficient attributes. Twenty members of the class were fitted with Caprotti valve-gear in place of the standard Walschaerts, and a number of those were based at Patricroft MPD. On 5 March 1965, No. 73134 was on shed. *(Opposite.)*

PLATE 11. Another of Patricroft's stud of BR/STD Class 5 4–6–0's. No. 73170 struggles to restart a heavy freight on a foggy, icy day – 1 December 1964. This scene on Patricroft triangle, of an engine now scrapped, on a line no longer in existence, shows a Lancashire long since gone.

Fig. 5 (page 43), Fig. 6 (page 44) and Fig. 7 (page 45).
The spawning railways of the early Victorian years reached out into the countryside, beyond the city of Manchester. Where railways went, suburbs followed. By the end of the nineteenth century the railway map of Lancashire looked like the meandering stagger of a drunken spider. Eccles, on the original (*continued.*)

More and more did they meet their Lancashire friends on their business visits to London!

Preston itself seems to have declined in dignity since the demise of steam. Proud Preston and Tom Finney were surely synonymous with First Division status for the steam railway. Diesels and electrics equate with relegation to the nether regions. Defoe in his *Itinerary* referred to Preston as 'this town the gentry resort in the winter for many miles around, and there are, during the season, assemblies, balls, etc'.

By 1797 an early trainroad linked two sections of the Lancaster Canal over a distance of five miles across the Ribble Valley between Walton and Preston. It crossed the Ribble by a bridge and tunnelled under part of Preston itself. The town soon became an important railway junction on the West Coast Main Line, with early railways incorporating the name of Preston within their title.

Cars and motorways to Blackpool have caused a dramatic decline both in rail commuter and excursion traffic through Preston. Although still an important station, the place lacks the excitement once associated with a main-line junction in the steam era.

Saturday 3 August 1968 saw arguably the last moment of real

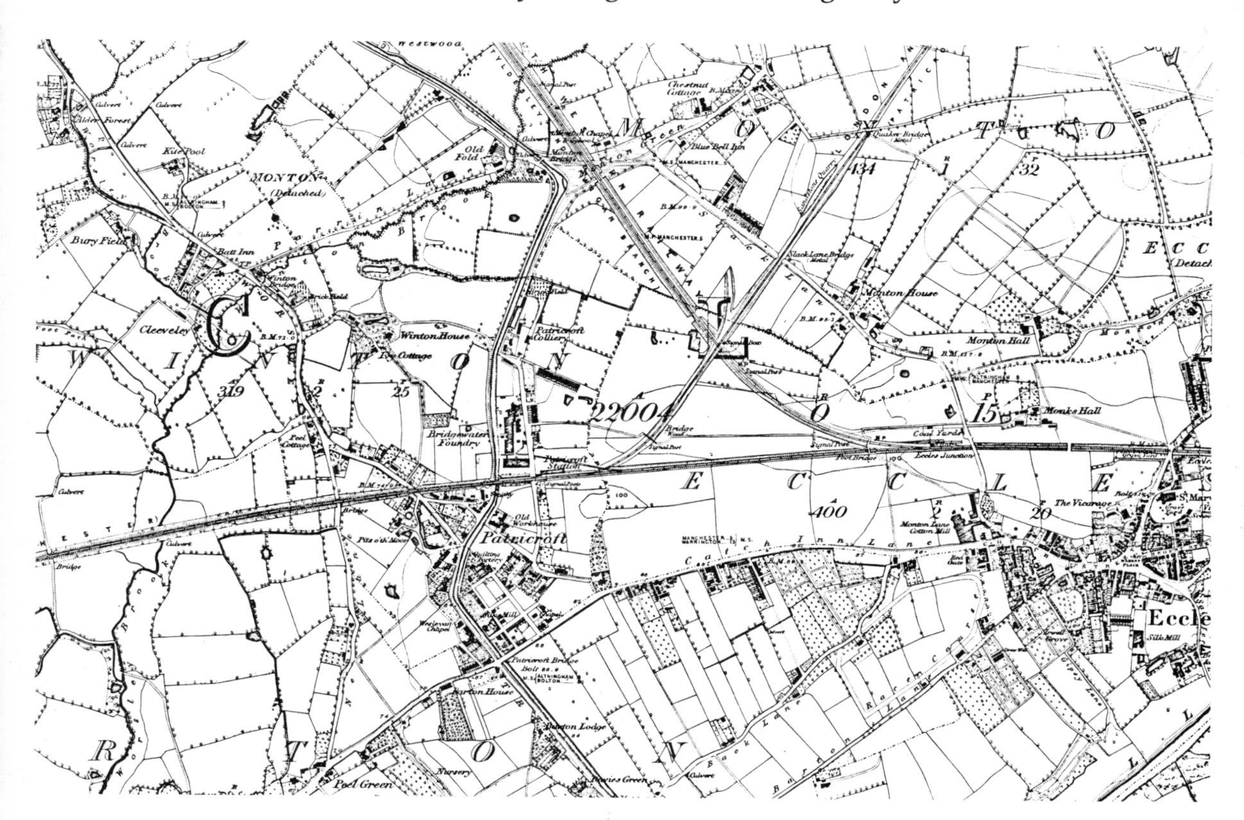

excitement at Preston station, when the final ordinary passenger train to be provided with steam haulage, the 21.25 Preston to Liverpool Exchange, departed on its historic journey. The station was thronged with people, and the few ordinary passengers must have wondered what on earth was happening. The engine that drew out of Preston that evening was one of Lostock Hall's Black 5s, No. 45318.

Lostock Hall and Rose Grove today give little clue of their once busy life. A steam engine shed was a living entity – men and machines both moving, breathing and talking; the eye, the ear and the nose ever alert to the sight, sound and smell of the world of steam.

The possession of a shed-pass to one of Lancashire's last

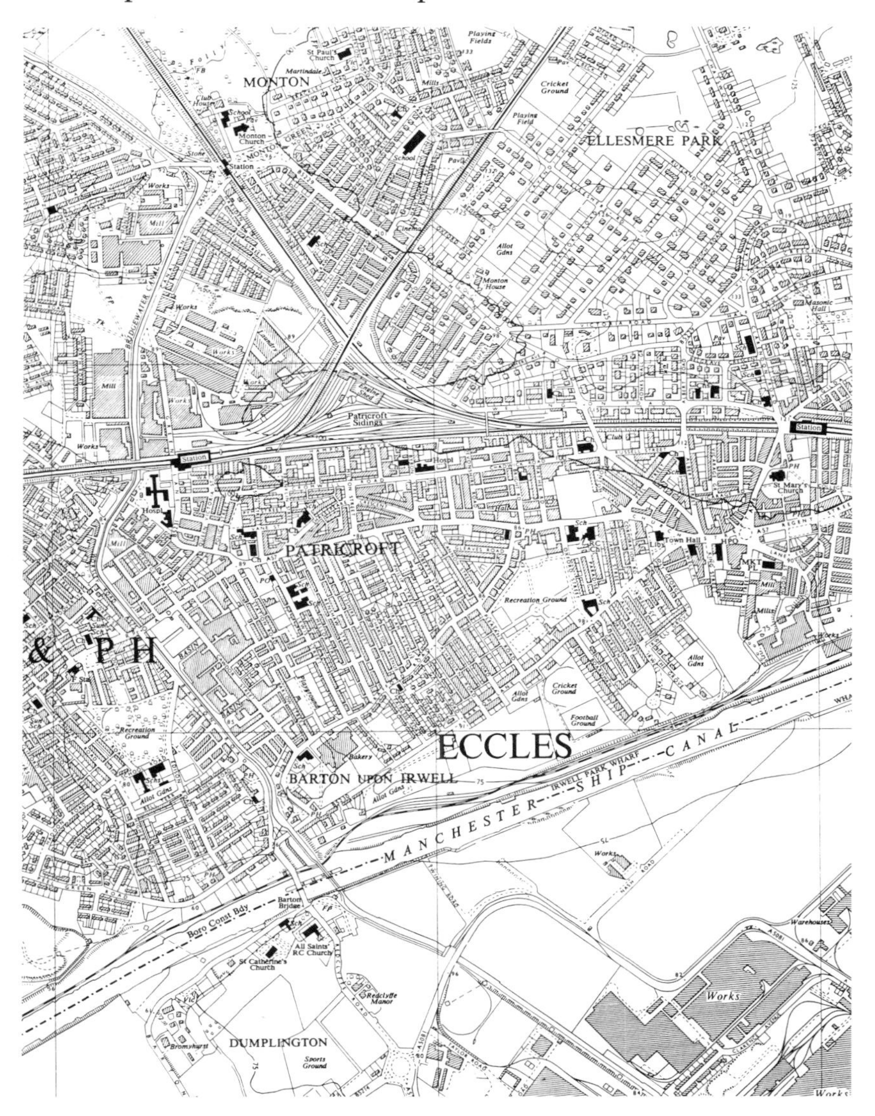

Liverpool-Manchester line, is associated with the first railway fatality, for it was here that the Duke of Wellington stopped to enquire about William Huskisson. The first of the two junctions from the L & M at Eccles ran from the east end of Patricroft Station, to Molyneaux Junction. Opened in 1850, it gave access to the East Lancashire Railway.
The second junction, half a mile east of Patricroft, was opened on 1 September 1864 by the LNWR. It ran to Springs Branch, Wigan. Thus was created Patricroft Triangle, within which was Patricroft Shed, with its turntable, visible on the 1954 map.
The 1977 map tells its own tale of decline. Gone are the junctions, track and railway buildings. No doubt ere long that distinctive white triangle of land will itself be subsumed under more unmemorable suburban architecture.

steam strongholds was a great prize to possess. Whilst trespassing was not unknown and whilst shed-masters rarely refused me entry to their sheds if I had no pass, I always preferred the tingling certainty of steam photography, to the tingling uncertainty of wondering if I was going to be allowed to roam at will around the shed and beyond its strict confines.

To disclose my tactics for gaining entry to sheds to which I had no permit might be to give away methods I may yet need to employ! Suffice to say that the interest in their job, their past, their present and their future shown towards railwaymen by enthusiasts, gave and gives a bond between us that survives even the demise of steam.

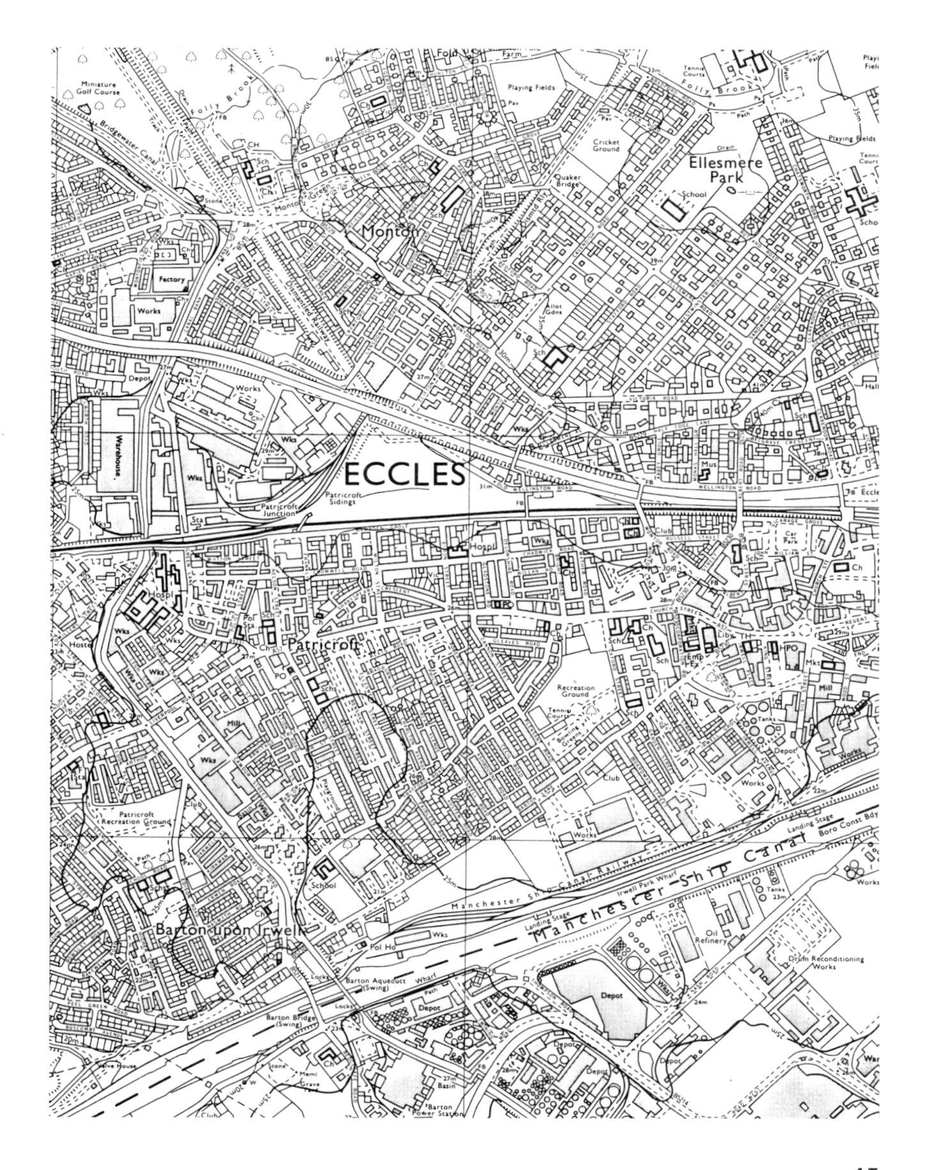

4615

3 Pannier Tanks

PLATE 12. *Moment typique.* A Class 57xx Pannier Tank, No. 4615, hustles a trip freight through Acton Wells Junction, on to the GW main line at Old Oak, as storm clouds gather on a bleak day in March 1963. *(Opposite.)*

PLATE 13. Colour photography of dirty engines at the end of the steam era, inside near-derelict sheds, without fast modern colour film, was a testing task for the railway photographer. On 5 July 1964, four GW Pannier Tanks were silhouetted against the light streaming into Old Oak Common MPD, following the demolition of one wall of the shed prior to reconstruction of the site of this most famous GWR steam shed. With apologies to Gilbert & Sullivan, I caption this shot, 'Four little girls from school are we . . .' Nos 4638 : 8459 : 9495 : 9470. *(Following pages.)*

No more descriptive phrase exists in the railway vocabulary than the words 'pannier tank'. Fussy as a – busy as a – self-important as a – . Try describing one of the Big Four in two words. The GWR lent itself most readily to personification, and the pannier tank, more than any other type of locomotive on any other railway, personified the GWR – even more than the copper-capped chimney, itself a feature which set the GWR apart. 'God's Wonderful Railway' was great indeed, and not least in its locomotive practice.

Numerous learned books recount the history of GWR motive power. Perhaps the GWR's outstanding feature was an uncanny ability to arrive at fundamentally sound solutions to its locomotive requirements, and to stick to them. Where the GWR ventured, others would surely follow; the great William Stanier was, after all, Swindon trained and bred. Loyalty to tried and trusted wheel arrangements remained an abiding feature of the GWR. From Churchward's 28xx, the first outside cylinder 2–8–0 on Britain's rails, flowed the most effective freight-engine wheel-arrangement Britain ever saw; through a series of 4–6–0 mixed-traffic designs which can claim to have sired Black 5s from Saints, Halls and Stars; down to the humble 0–6–0 tank engine.

I know not if anyone has written a history of the pannier tank. That is not my purpose – I come to salute the pannier, not to chronicle it. First introduced in the late-nineteenth century, the design survived amalgamations, grouping and nationalisation. Pannier tanks were still being built at Swindon in the early 1950s, to a design and in a style that Churchward would undoubtedly have recognised. Indeed the 16xx class was not introduced by Swindon until 1949, after nationalisation, and they were still being built as recently as 1955.

Potted histories of Britain's railways always seem to concentrate on the glamorous engines; yet as modesty is not a disease from which I suffer, it seems to me that the success of my

first railway book, *British Steam in Cameracolour*, was because of, and not in spite of, the fact it portrayed the declining years of working steam in Britain in their grubby, unglamorous reality. The debt owed by legions of railway enthusiasts to the pioneers of the preservation movement will remain forever. The phrase 'stuffed steam' to describe the gleaming *Mallard, KGv, Flying Scotsman* or *Kholapur* is not intended to denigrate the efforts of the preservationists, but to recognise the quite separate appeal of the grimy workaday tank engine shunting in Hackney yard at Newton Abbot, from the slightly ethereal dance of an A4 down Stoke Bank. Thus is the separation created between the hard-core railway enthusiast and the chap who likes a pretty picture. Pannier tanks working out their last few years on the Western Region were not the obvious magnet for photographers; and who remembers many photographs of panniers in Sonning Cutting?

The 'standard' GWR pannier tank of 57xx class, was indeed the most numerous single class of steam engine ever built in Britain: 863 were constructed between 1929 and 1950.

They were the last class of GWR engine to survive the final massacre of GWR steam. Twenty-seven 57xx, plus three 16xx, saw the first light of 1966 – by the year's end they had all gone. With them went the last vestige of GWR motive power; after that it was merely Western Region. As work and effort seemed more in keeping with cold and rain than with sparkling sunshine, so my record of the decline and fall of the pannier tank seemed to be linked to poor photographic conditions. If I were forced to choose the one single picture that epitomised the transplantation of steam by diesel, I would select the sight of four pannier tanks silhouetted in the round house at Old Oak Common. They are standing gaunt against light let into the shed by the demolition of one wall; part of the destruction of the old shed prior to the building of the new diesel depot.

Not everybody loved the panniers – because not everybody loved the GWR! For more than a century, fierce rivalry existed in the West Country between the Great Western and the London and South Western Railways. The deep-seated loyalties, and the equally deep-seated jealousies, between the two were not removed by the 1923 grouping. The Southern Railway as successor to the LSWR fought hard to maintain and improve its service to Exeter, Plymouth and beyond. When the bureaucrats finally got at the structural organisation of BR, and created areas of jurisdiction based on neatness rather than history, the first act of the usurper was to insert his motive power at the expense of the traditional engines. At towns like Yeovil, formerly meeting places between GWR and SR, the Western Region swept away the LSWR M7 tanks on the Yeovil Junction/Yeovil Town service

PLATE 14. One of the last regular passenger turns on which Panniers were employed: Class 57xx No. 4631 blasts away from Highbridge S&D with the 16.00 to Evercreech Junction on 20 February 1965. Many S&D men deeply resented the replacement of LMS engines by GWR motive power and claimed that the Panniers battered the track across the Sedgemoor peats.

at the earliest opportunity and the pannier tank was brought on to the scene. They were also 'imported' onto the Somerset & Dorset in that line's death-throes, to replace the S & D's Midland-designed tank engines of vintage years. (See Chapter 12.)

Some of the odder-looking pannier tanks were those modified to condensing engines to work over the lines in the tunnels of the Metropolitan Railway, to Smithfield. After the initial experimental conversion, ten of these modified engines were built at Swindon in 1933, and I was able to photograph one at Olympia in November 1962. These engines combined pannier with side-tanks and had an increased water-capacity of 1,230 gallons. They had to conform to London Transport regulations for working over the electrified lines.

The tried and trusted pannier tank became the last engine to be built in any number by British Railways that was totally identified with a pre-nationalisation design. In 1947 Hawksworth introduced his 94xx class 0–6–0 PT heavy shunting engine. Ten were built at Swindon in that year, the last year of the GWR. Thereafter a further two hundred engines were built for BR between 1950 and 1956 by outside contractors. In terms of their life-expectancy, they were really a waste of money, because BR were already examining the possibility of changing from steam to

3709

PLATE 15. Pannier Requiem. With steam's total eclipse on the Western Region by the end of 1965, the only sheds still with active GWR engines in 1966 were in those areas in North Wales and the North-West which had been transferred from Western to Midland Region administration. As the MR's policy was to condemn their WR stock as soon as possible, Croes Newydd presented one of the final scenes of Panniers in steam albeit in their death-throes when I visited the shed in March 1966 and photographed 57xx class engine; No. 3709 is still alive.

diesel for shunting duties. The 94xxs were withdrawn with years of useful life left in them; the first to be withdrawn, No. 8417, was scrapped in 1959, a mere three years after the last of the class had entered service.

In 1949 Hawksworth produced ten engines of Class 15xx, another 0–6–0 pannier tank for heavy shunting duties. They were the heaviest pannier tanks built for the Great Western, although, of course, being BR-built they were not officially GWR engines; yet nobody could mistake their heritage. They too were too late to make a significant impact on future motive-power developments.

The last of this trio of post nationalisation panniers – and Hawksworth's last locomotive design for the GWR – was the 16xx class of light shunting engines. In retrospect, it is surprising that these chirpy little engines ever experienced the feel of metal under their wheels. With the emergence, in May 1955, of No. 1669 from Swindon came the end of a long long line of pannier tanks built there. Thereafter Swindon turned out only BR standard engines. One of the class, No. 1628, was amongst the last thirty engines withdrawn at their end of steam in 1966.

·30315·
30315

4 Charlies, Crabs and Woolworths

Steam engines differ from other mechanical devices because they share many of man's natural attributes. Unlike a diesel or electric locomotive, you cannot simply switch them on or off. They are either alive, ill or dead – the latter state generally indicating the need for hospitalisation (overhaul) or mortuary (scrapping).

In fact, identifying the steam engine with *homo sapiens* is somewhat unfair to the former. Reliability, willingness to tackle unaccustomed tasks and never taking industrial action are features that endeared an engine to her shedmaster, and accounted for the cosy atmosphere of familiarity between steam engines and railwaymen that can never be replaced by modern motive power counterparts. That is why we gave them nicknames.

Nicknames are not always terms of endearment; sometimes they are more descriptive than fraternal. Yet relating a nickname to a whole class of engine rather than to one individual, is indicative of our relationship with the steam locomotive.

This is a dangerous chapter! The source of many nicknames is a matter of hot dispute amongst railwaymen and enthusiasts. Who first called a Crab a 'Crab'? Why was a Q1 a 'Charlie'? What about 'Black Motors', 'Flashers' and 'Camels'? To detail all known engine nicknames requires a book to itself. Perhaps the publishers will invite me to prepare just such a learned tome! 'Austin 7', 'Baby Scot', 'Jinty', 'Black-un', 'Mickey-Mouse' – the list seems endless. Some were prosaic, some affectionate, some downright rude.

My purpose here is not educational. Perhaps it is not too difficult to attach the appropriate caption to the relevant picture – but as competitions are in vogue, there will be no prizes for selecting which is which!

The Crab has always been one of my favourite engines. Although the class was introduced in 1926, some three years after the grouping which created the LMS, and some four years after the merger between the Lancashire & Yorkshire Railway (L&YR),

PLATE 16. A Black Motor – a rare colour photograph of a long-lived class of engine introduced by the LSWR in 1897. This is No. 30315, at Eastleigh, on 18 May 1963, awaiting cutting up. *(Opposite.)*

PLATE 17. LMS Hughes-Fowler 2–6–0 No. 42760 at rest inside Nottingham Midland MPD on 16 July 1964. It was withdrawn the following month. *(Following pages.)*

and the London and North Western Railway (LNWR), the Crab has a definite Lancashire & Yorkshire appearance. As is often the case, the chimney is one of the main distinguishing features, and the Crab chimney has the Horwich hallmark firmly stamped upon it.

PLATE 18. Charlie scrubbed! 'Q1' 0–6–0 No. 33026 rests on Feltham shed on 3 October 1964, prior to working an enthusiasts special the next day. *(Above.)*

The Lancashire & Yorkshire has a fascination for me, because of the time I spent photographing steam in Lancashire in the mid 1960s. This railway fired my imagination as perhaps the last identifiable survivor of the nineteenth-century second-rank railways. As a Southerner born and bred, I have always admired the men, methods and mechanics of this northern English railway, which seemed to retain intimate links with its own operating environment. No distant London-based tycoons issued instructions to the men of the Lancashire & Yorkshire.

As I identify the L & Y with the brawn and native wit of Lancashire and Yorkshire, so I identify the Crabs with hard work. In my mind's eye live the photographs by Derek Cross of Crabs at Dailly in Ayrshire working the Bargany coal-trains. The class nickname stems from the angle of the cylinders; functional rather than attractive. The high running-plate predated by twenty-one years Ivatt's LMS mogul, which itself earned the unlovely but not undeserved nickname of 'Flying Pig'. With the running-plate so

PLATE 19. Crab motion: the nickname was born from the angle of the cylinders. No. 42777 was photographed at Birkenhead on 21 August 1965, some 38 years to the month after she was built. The last Crab in service, No. 42942, was withdrawn from Birkenhead MPD in January 1967. *(Opposite.)*

PLATE 20. Charlie's interior! Bulleid 'Q1' 0–6–0 No. 33018 undergoes routine maintenance at Three Bridges in April 1963.

high up on the boiler flanks, the wheels and frames were completely exposed. Splashers are to driving-wheels what petticoats are to thighs; total exposure is a lot less attractive than a well-designed encasement. The inherent ugliness of Ivatt's Flying Pigs was completed by the initial attachment of a huge double chimney of most unattractive shape, which seemed grotesquely large.

What Ivatt's 'Flying Pigs lacked in aesthetics they gained in reputation. Their performance justified their redesignation from their original conception of 4F replacement for the time-honoured Midland 0–6–0s, to 4P/4F and thence 4MT in BR days.

Subsequently the double chimney was replaced by a more acceptable single chimney. The design was perpetuated in the British Rail Standard 'Class 4' mixed-traffic engines designed at Doncaster and introduced in 1953, although some of the roughest edges of the Flying Pigs were mitigated; for example, the raised footplating was linked to the buffer-beam by a drop-end. But the 76xxx engines had little charm, having replaced the Ivatt ugliness with the inevitably nondescript appearance of engines more the product of a committee than the inborn imprint of the Chief Mechanical Engineer; Ivatt himself being the last of the four CMEs to go following nationalisation in 1948.

If character and individuality were inimical with nationalisation, it is fortunate indeed that Bulleid and BR never tried to co-exist! In 1951, at the age of 70, he became CME of the Irish National Rail System. OVS Bulleid has left as large an imprint in the minds of Britain's railway historians as any of his great predecessors. Taking over as CME of the Southern Railway in 1937, his impact on SR locomotive practice and performance only reached maturity coincidentally with World War 2. His first freight design, the Q1, quickly acquired the reputation as the ugliest engine ever designed and built for Britain's railway system! Like the Flying Pigs, the Charlies – as the QIs became known – were designed around ease of maintenance. Even in wartime enthusiasts were shocked with Bulleid's apparently total disregard for the appearance of his engines.

The Charlies were most successful engines. When new, the pioneer engine of the class underwent a trial with a train of 1,000 tons, running the 24 miles from Woking to Basingstoke with ease in eight minutes less than the scheduled time for an 800 ton train on this section. Thus the Charlies proved again that good looks were not a necessary concomitant to good performance.

Appearance was not the only procreator of nicknames, although I always smile at the epithet of 'Hikers' given to the Great Eastern B12s fitted with ACFI water-feed heaters, the contraption providing instantly visible justification for the title. Alas my tender youthfulness precluded my ever photographing one of these beasts. I was more fortunate with the Woolworths, the appearance of which had nothing whatsoever to do with their nickname.

If the epithet 'Woolworth' implied fulfilling the varied requirements of busy people within one 'shopfront' then Maunsell's N class 2–6–0s, the prototype of which was built at Ashford in 1917, well earned that title. Most of the Ns saw more than forty years service, and the demise of steam, not their condition, brought about their withdrawal. However, the nickname derives not from their virtues but from the fact that

PLATE 21. Another Crab – performing the task for which it was designed. The Crab was an 'Engineman's Engine' – not glamorised by the LMS who introduced it in 1926. Nearly forty years later, this Hughes/Fowler mogul, No. 42892, wheels a heavy freight southwards past Warrington Dallam MPD on a misty day, 2 December 1964. This is one of my favourite photographs. *(Following pages.)*

PLATE 22. A Woolworth at work. Reliable, sure-footed and undramatic, No. 31871 heads the 08.45 Margate–Wolverhampton through-train between Dorking Town and Gomshall and Shere, on 20 July 1963.

fifty Ns were commissioned by the Government in 1920/22 and built in kit form at Woolwich Arsenal to alleviate unemployment. Woolwich was soon corrupted to Woolworth, although the originator of this neat transposition remains anonymous. In fact, the engines built at Woolwich were purchased by the Southern Railway – at a bargain price – in 1924. Subsequently Ashford turned out fifteen more of these most useful moguls: added to the fifteen built pre-Woolwich and you have a class of eighty engines that saw service on SR across that Railway's territory from East Kent to North Cornwall.

Ironically, there is a link between the Woolworths and the Charlies that few perhaps have noticed. Whereas Bulleid went from his job as CME of the SR to take over the CIE in Ireland, Maunsell came from Ireland to become CME of the South Eastern & Chatham Railway (SECR) and thereafter of the Southern. This 'Irish Connection' was completed by the purchase of twenty-seven Woolwich kits by Irish Railways; engines designed by Maunsell, formerly locomotive superintendent of the Great Southern & Western Railway, 1911–13. Strangely, only twenty-six engines were completed in Ireland from the twenty-seven kits purchased. I suppose that was better, if less in character, than if twenty-eight engines had been built from twenty-seven kits.

But – I digress! The Woolworths gave the SR and latterly Southern Region of BR, years of reliable service. Perhaps one of the first precursors of the standard engine, they were popular with footplate-crews and shed-staff alike, being not only reliable but low on maintenance costs. Their swansong was in passenger service on the former SECR line, Tonbridge/Redhill/Guildford/Reading. (See Chapter 9.) As late as November 1964 I recall the pleasure of photographing No. 31873 at Reading (SR) MPD. A genuine Woolworth, No. 31873, entered service in September 1925 and, with 31866, was one of the only two that survived in service into 1966.

Whereas the name 'Woolworth' has to do with history, 'Flying Pig' with design, 'Charlie' with character, the final words I dedicate to Drummond's London & South Western Railway T9s – the 'Greyhounds'. Introduced in 1899, these free-running stylish 4–4–0s were splendid engines.

Eleven lasted into 1961, and some amassed around two million miles of revenue-earning service. To those who know and love North Cornwall, to whom the words Bodmin and Wadebridge imply more than an Emmett-generated August traffic jam, the passing of the Greyhounds and indeed the Woolworths from the 'Withered Arm' signalled the end of an era. No picture of a daintily-preserved engine can do justice to that engine in its prime, or in its context. But I have allowed myself the intrusion of restored Greyhound No. 120 at Eastleigh, to indicate the aptness of her nickname: (See Plate 33, Chapter 7.)

48393

5 Last Weeks of Steam

PLATE 23. The last weeks of steam were concentrated mainly on freight services – much of it on the former LMS lines in industrial Lancashire. Between Blackburn and Preston, however, is a small stretch of open countryside which provided an agreeable backdrop for photographing the neglected steam engines, whose working life was terminated on Sunday 4 August 1968. In this photograph, taken four days previously, Stainer '8F' 2–8–0 No. 48393 wheels a Rose Grove to Wyre Dock freight past Mintholme crossing. *(Opposite.)*

PLATE 24. Later the same afternoon, another freight makes its way west behind '8F' No. 48665, seen plodding past the crossing at Mintholme. *(Following pages.)*

It is nearly fifteen years since I finally slunk silently away from Rose Grove engine shed at Burnley, that Wednesday night. The day had been glorious; warm summer sunshine interspersed with fleeting cloud, with never a hint of rain. It was the last week of steam operation on Britain's railways. That era ushered in by Rocket at Rainhill was fading away, some few miles north in industrial Lancashire. Like some revered relative visited in an oriental death-house, my final respects had been paid.

Three days later, hundreds, indeed thousands, would spend that last Saturday swarming like bees around the last three steam-worked sheds: I wanted my last day of steam to be a working weekday – a day as near to normality as would be lived and remembered. I preferred to pay my last respects to the colossus that was the working steam engine, alone. Never shall I forget the pangs of hopelessness that swept across me as, kicking the Rose Grove ash from my shoes, I watched the sun cast its fading glow on a scene that I should never see again . . .

Everyone knew that steam's final scheduled working day was Saturday, 3 August 1968. Oh yes, I know that there were 'steam farewells' scheduled for the Sunday, and indeed one the next weekend. But these were 'exhibitions'; worthy, but for me, macabre. I did not regard the end of steam as an occasion to celebrate behind some tarted-up 'Britannia'. As for drinking champagne on the 15 guinea a head 'Last Steam Train' excursion – that seemed like knees-up-Mother Brown at a State funeral.

My increasingly frequent forays to Preston, Burnley and Carnforth in June and July 1968 caused little comment to my commercial masters in London. Who would believe that visits to the North-West by an executive of a company with interests in hotels, travel, transport and the motor industry were undertaken other than in the line of duty? Yet for me, it was impossible to spend a more enjoyable day than photographing working steam, on shed or in the Lancashire landscape; then ending the day with

48665

PLATE 25. Four days to go. and amongst the last survivors is one of only 4 out of 842 Stanier Black 5s that were named. No. 45156 *Ayrshire Yeomanry* was built by Armstrong Whitworth in 1935. I photographed her on 31 July 1968, as she approached Rose Grove station.

an excellent dinner at the French Restaurant at Manchester's Midland Hotel – owned by British Rail, of course!

Lunchtime visits to Rose Grove and Lostock Hall grew longer as the summer of 1968 slipped by. My smart black shoes were often covered in oil, grease and ash as I rushed to keep my afternoon business appointments. My city suit and stiff white collar must have looked singularly inappropriate as I squatted in one of Lostock Hall's ashpits to record a passing Black 5.

Between Preston and Burnley lies much of industrial Lancashire – and some green fields too. Indeed, pictures of freight-trains with captions like 'near Hoghton Bottom' give little clue to the location, to those unfamiliar with the terrain.

Working from my London office severely limited the time I could spend in Lancashire: yet, such ingenuity as I possessed, that summer of '68, was bent towards devising visits north. The last sheds with an allocation of steam motive power in the Manchester Division of the London Midland Region closed on 1 July.

The demise of these three sheds, Newton Heath, Patricroft and Bolton, left only Carnforth, Lostock Hall and Rose Grove with an active steam allocation. The closure of the three Manchester area sheds had two byproducts; firstly, the tightening vice on the steam era brought an increasingly frantic group of enthusiasts ever more closely into each other's company; secondly, it ensured that the steam finale in Britain would be in the hands, not of post-nationalisation-designed and built engines, but in the most worthy and deserving care of Sir William Stanier's Black 5s and 8Fs. Notwithstanding my love for and loyalty to the GWR, it was wholly appropriate, given that Lancashire was to witness steam's 'Final Act', that the LMS, not BR, should provide the motive power.

I cannot, nor do I intend to try to, pretend that I was not emotional about the end of steam. I preferred to be alone, those final weeks and days. Not for me casual relationships struck up with other enthusiasts at Rose Grove and Lostock Hall, places which became like gold-prospectors' haunts those last few weeks. Around every engine were fellow ghouls, some with cameras, some with tape-recorders and some with rags. The extraordinary sight of steam enthusiasts clearing away encrusted grime and grease from dirty, filthy neglected engines, in order to ensure that the last few working days were spent looking as though someone cared for them, was a more eloquent testimony to man's love of steam than any words of mine.

I was not at Lostock Hall on Sunday, 4 August. The shed was a rare sight that morning, so they say. Never before have so many rail tours run on one day, as locomotives went off shed at intervals up to lunchtime. Doubtless those engines could still be serving British Rail efficiently, burning British coal, if it had not been decreed otherwise.

Perhaps it is fortunate for us railway enthusiasts that the news-media, be it *Railway World, Railway Magazine* or *Railway Observer*, are prosaic rather than flamboyant. How phlegmatic is the following extract from *The Railway Observer*, **Vol 38, No. 476, October 1968.**

WITHDRAWALS – 4–6–2 Cl.7 70013 Oliver Cromwell † (10A) 4–6–0 Cl.5 44781 (10A) 44871 (10A) 45110† (10D). Diesel-mechanical Class 4 D2213 (8H), D2218 (5A). Electric Class 81 E3009.

† *Class Extinct. The withdrawal in w/e 17th August of 70013 'Oliver Cromwell' and the three Stanier Cl.5s rendered the standard-gauge steam locomotive extinct on BR. Cl.8F 48448 should be deleted from last month's note on the last surviving steam locomotives: it is correctly shewn on the same page as withdrawn P.8/68.*

The details of the last steam-hauled passenger runs on the Preston/Liverpool and Preston/Blackpool trains are well-enough known to real enthusiasts not to need repeating here. Most railway books, however, are either photographic with captions, or learned historic accounts of particular lines, designs or times. Perhaps I may conclude this personal testimony to the last scheduled regular steam workings on BR, by detailing, as a matter of fact, the numbers and date of construction of the final batch of Black 5 withdrawals from Carnforth, Lostock Hall and Rose Grove.

44690	Horwich	1950
44709	Horwich	1948
44713	Horwich	1948
44735	Crewe	1949
44781	Crewe	1949
44806	Derby	1944
44809	Derby	1944
44871	Crewe	1944/6
44874	Crewe	1944/6
44877	Crewe	1944/6
44888	Crewe	1944/6
44894	Crewe	1944/6
44897	Crewe	1944/6
44932	Horwich	1945/6
44950	Horwich	1945/6
44971	Crewe	1946
45017	Crewe	1935
45025	Vulcan Foundry	1934/5
45055	Vulcan Foundry	1934/5
45073	Crewe	1935
45095	Vulcan Foundry	1935
45096	Vulcan Foundry	1935
45110	Vulcan Foundry	1935
45134	Armstrong Whitworth	1935
45156	Armstrong Whitworth	1935
45206	Armstrong Whitworth	1935
45212	Armstrong Whitworth	1935
45231	Armstrong Whitworth	1936/7
45260	Armstrong Whitworth	1936/7
45262	Armstrong Whitworth	1936/7
45268	Armstrong Whitworth	1936/7
45269	Armstrong Whitworth	1936/7
45287	Armstrong Whitworth	1936/7
45305	Armstrong Whitworth	1936/7
45310	Armstrong Whitworth	1936/7
45318	Armstrong Whitworth	1936/7
45330	Armstrong Whitworth	1936/7
45342	Armstrong Whitworth	1936/7
45350	Armstrong Whitworth	1936/7
45386	Armstrong Whitworth	1936/7

45388	Armstrong Whitworth	1936/7
45390	Armstrong Whitworth	1936/7
45397	Armstrong Whitworth	1936/7
45407	Armstrong Whitworth	1936/7
45444	Armstrong Whitworth	1936/7
45447	Armstrong Whitworth	1936/7

The fact that so many pre-war Black 5s survived to the bitter end, speaks for itself.

Ode to the passing of the steam era from Britain's three last steam-engine sheds.

Incongruous it was, at Lostock Hall
Even to contemplate steam's decline and fall
Everything normal – coaling and taking water
Black Fives at work: not lambs – nor for the slaughter . . .

July 'Sixty Eight – summer as sultry as ever
Rose Grove at Burnley – 8Fs with noble endeavour
On coal trains from Yorkshire: imagine them not hauled by steam?
Its unthinkable – 'nowt but a terrible dream'.

At desolate Carnforth – of Furness and London North Western
The shed is its centre: its closure a wretched suggestion.
With history eternally linked to nothing but rail
An engine to Carnforth relates like a yacht to a sail

Yet these are the sheds that fate has decreed will endure
The mourners: whose closure extinguishes that allure
Of the steam-engine: servant and friend
Whose passing has caused us to wend
Our painstaking way, depressingly one to the other
In desperate search: each enthusiast, like a lost brother
Embraces an engine with camera. An era is gone:
From 'Rocket' to Riddles the memories merge into one.

And now, as we contemplate dreariness – life without steam
We have little left save a life that must certainly seem
As empty as Carnforth, as Rose Grove and as Lostock Hall
Without Stanier's whistles – stillness: death's pall . . .

A03

6 Berks & Hants

PLATE 26. BR-built double-chimney 'Castle' Class 4–6–0 No. 7002 *Devizes Castle* passes through Sonning Cutting on 22 June 1963 with the 13.15 down express from Paddington to Worcester and Hereford. This was the last express passenger service from Paddington to remain regularly steam hauled. *(Opposite.)*

PLATE 27. By 7 November 1964 this sight was distinctly uncommon. Excepting the Worcester services, most passenger and van trains were diesel-hauled, whilst steam-haulage, largely in the hands of Halls, was restricted to freight trains or used as a substitute for defective diesels. The sight of a GWR 28xx 2–8–0 hurtling eastwards, near Reading West Junction, in winter light, caused me to attempt to pan this shot, and it didn't quite work! No. 3859 has charge of the 08.00 Cardiff to Old Oak Empty Vans Special.

PLATE 28. Sunshine in late September, on the 25th of the month, 1963, adds sparkle to briskness as '9F' 2–10–0 No. 92001 hustles Esso tanks from Fawley up the SW main line east of Winchfield. Steam-hauled express freight and passenger services were a fine sight at this point, speeding along the embankment, especially when viewed from the pastures to the south of the line. *(Following pages.)*

For those who turn with fumbling excitement to this chapter, hoping to see sparkling Kings rushing through Newbury with the *Cornish Riviera* – I'm sorry! To be honest, this chapter is an afterthought born from the absence of an appropriate location for some of the shots I wanted to include in this book. Previously I had decided that Chapter 1 could be extended, by means of 'author's licence' to include photographs of the last lap of the journey up the LSWR and GWR main lines to Waterloo and Paddington respectively.

'Basingstoke is really London' I told Jane. 'It may be now' she replied – and indeed it is – 'but in the days of steam it was still in Hampshire!'

Reading, I pointed out, was the start of the 'last lap' to London.

'You can't get away with that either' said the conscience that passes for a wife!

The fact is that when we lived in Sunningdale, either Reading or Basingstoke made a not too-distant journey. With two main lines, including Sonning and Winchfield cuttings, as well as the Basingstoke to Reading link that saw the final runs of the Pines Express, the two counties provide sufficient scope for interesting photography to warrant the inclusion of this short chapter.

PLATE 29. With Kings, Castles, Halls and Manors preserved, the 'Grange' Class 4–6–0s of the GWR alone are totally extinct from the GWR engines of this wheel-arrangement in service in the 1960s. No. 6855 *Saighton Grange*, was immaculate at Reading MPD on 4 May 1963; she was withdrawn in October 1965. Reading GWR shed always had a stud of mixed-traffic engines. Whilst Granges attracted little attention, they rarely let the operating department down.

7 Specials

In the world of make-believe, all kings, queens, chiefs and rulers are either loyal ministering angels or wicked loathsome devils. In the real world, the angels and devils exist within the general run of mankind, sandwiched between the teeming millions of ordinary folk. When my publisher asked me to write a chapter on special trains he, an ordinary mortal not a railway enthusiast, fell into the casual language used by those sad folk who do not *believe* in railways as a way of life, and his language showed that he felt 'pretty' trains were 'interesting' trains. If indeed 'twere so, then precious few books on British Rail's declining years would have reached the bookshops, for gleaming paintwork was not a feature of steam in the 1960s. Indeed steam passed through Shakespeare's 'Seven Ages of Man', and those last six years encapsulated in my title were a reminder that the

'Last Scene of all,
That Ends this strange eventful history,
Is second childishness and mere oblivion,
Sans teeth, sans eyes, sans taste, sans everything'
(As You Like It)

In steam's dotage, grime and decline went hand in hand towards the grave, death being 3 August 1968 and burial some eight days later. If preservation be reincarnation, then steam lives on; yet if truth be told, the gleaming paintwork of 'stuffed steam' is a pale substitute for a living, natural working steam railway.

But, back to my publisher! 'Give me specials' he intones. 'Royal Trains, the *Golden Arrow*, you know the sort of thing I mean'. Indeed Sir I do, I remember them well; recorded faithfully by Bishop Treacy and the photographers of yesteryear – (Dear Bishop, is there an LNER in heaven?) By 1963 steam's role on titled trains was almost a memory, and my book is about the real world. The Lords and Masters of our railway system, faced with the increasingly rare arrival of a foreign potentate by rail rather

PLATE 30. 'Special' in every way; unrebuilt 'Battle of Britain' 4–6–2 No. 34051 *Winston Churchill* hauls the great man's all-Pullman funeral train from Waterloo to Bladon, seen here near Sunningdale. In the pushchair is our elder son Simon, then aged 11 months. *(Following pages.)*

than by air, would turn out a dreadful diesel or a lifeless electric locomotive to heave the visitor to the capital.

Yet for a handful of great men, not even a steam engine would match their contribution to our nation's history. Who would design or dedicate an engine to Sir Winston? Step forward, the engineer who thus dares. Inventive genius, challenging assumptions, fearless of critics, willing to attempt the unthinkable, a man whose willingness to learn from history arms him not with repetition but with adventure – O V S Bulleid.

When Churchill died, there stirred in the breasts of Southern Region, the feeling that steam should haul the old warrior to his grave. There was not far to turn. 'Battle of Britain'. Unrebuilt. Pacific. No. 34051 *Winston Churchill*. Along suburban embankments they stood in silence as the warrior passed. No one would see the like of him again.

From the last of the great warriors it is but a short step to the last 'King'. Epitomising Britain's power, dignity and glory, it was naturally the Great Western Railway, in 1927, that introduced the engine with the greatest Tractive Effort in the land. Built between 1927 and 1930, to haul the trains on Britain's longest non-stop run, London to Plymouth, the Kings retained their air of massive dignity to their demise in 1962. Of the final four officially withdrawn, 6018 *King Henry VI* was retained to work a Stephenson Locomotive Society (SLS) special on 28 April 1963.

As on the occasion of Sir Winston Churchill's funeral train, people came from far and wide to see *King Henry VI* on the SLS Special. At Southall I used my shed-pass to walk on to the Great Western main line to see her. Expectantly enjoying her approach on the down main line, I nearly met my maker as a d.m.u. sped past on the adjacent track. Not a 'Royal Train' in your interpretation dear Publisher, nevertheless, the passing of a King!

It is a long way from kings, and even further from royal trains, to today's 'football specials'. In truth it would be hard to design stock of sufficient discomfort to be suitable for the occupants of many of these trains. However, 15 years ago, things were not so bad. If a northern team were in the cup final then a stream of steam-hauled Specials would trundle up to Euston or Kings Cross, St Pancras or Marylebone, so a visit to these lines repaid the effort of being in London on that Saturday afternoon. Often the return journey north presented an even better spectacle, with the long late-spring evenings providing warm light for colour photography.

One of the last occasions on which the 'Premier Line' out of Euston saw a trail of these steam-hauled football specials returning north, was the occasion of the England v Scotland

PLATE 31. Football Special. 'Britannia' 4–6–2 No. 70026 *Polar Star* brings a trainload of Scots 'to town' for the England v Scotland international on 6 April 1963; seen here at Wembley Central *en route* for Euston. These football specials produced varied motive power, often from distant sheds.

PLATE 32. My nearest date with destiny came on 28 April 1963 – and it served me right! On that date, No. 6018 *King Henry VI*, officially withdrawn the previous December, worked an SLS farewell special, seen here at Southall. My enthusiasm to record a 'King' at work led me to abuse my Southall shed-pass and abandon my normal caution, much to the anger, I imagine, of the driver of the d.m.u. on whose track I was standing. These few seconds spent between the d.m.u and the special were not pleasant!

International at Wembley stadium on 6 April 1963. Stationed just north of Wembley Central station one was treated to a feast of steam; added to which was the additional pleasure provided by some vintage LMS stock, such as was provided for those fearsome 'schools specials' on which I travelled back to Uppingham from Euston thrice-yearly between 1948 and 1952.

From school trains to 'Schools' and Uppingham rates another mention. Its stance as the only school to reject the naming of an engine as a member of the class, has always caused me Freudian twitches! Originally 30923 was named *Uppingham* by Maunsell and the SR team. However, in the face of objection by the Headmaster, the name was changed to *Bradfield*. What a demotion for an engine.

The reporting code and presence of an engine I photographed at Eastleigh on 23 April 1963 qualifies it for a mention in this chapter. The last remaining 'Schools' class engines were withdrawn in December 1962. Most were hauled to Eastleigh for cutting-up. Visiting Basingstoke shed on 18 April 1963 I was astonished to see 30934 *St Lawrence* in steam inside the shed. My shot of the engine was hampered by cramped conditions and poor light.

Later that same day, *St Lawrence* was on shed at Eastleigh in a more photogenic location. My photograph of her in steam may be the last photograph ever taken of that class in steam in a regular service environment – a 'Special' indeed. Often have I wondered why she was steamed, rather than hauled dead for the journey from Basingstoke. I enjoy the thought that someone, somewhere, was determined to ensure she made her final trip with dignity.

For 'Preservation Specials', the 1960s were unique. The end of steam was nigh, and preparations were afoot to save as much as

PLATE 33. On 18 May 1963, LSWR T9 4–4–0 was on shed at Eastleigh. Now in the National Railway Museum's collection at York, she was built at Nine Elms in 1899 and remained in service with LSWR, SR and BR to become the last 'Greyhound' in traffic on withdrawal in 1963, when she was repainted in LSWR colours, as seen here. (In her LSWR days she never actually ran in this particular condition.) *(Opposite.)*

possible from the cutter's torch. Yet as long as steam was still in regular service there was the chance to see a gleaming, restored special in the working railway environment. Merchant Navies went North: A4s went West: and 4472 *Flying Scotsman* seemed to go everywhere!

On 18 May 1963, the Gainsborough Model Railway Society chartered 4472 for a run from Lincoln to Eastleigh. She took water at Basingstoke, and was coaled and fuelled at Eastleigh. The incongruity of a gleaming A3 at the home of the LSWR was heightened by the sight of her standing outside the shed alongside a humbler, and indeed much older engine, an LSWR M7, number 30251 from a class introduced in 1897. The Drummond M7s, many of which lasted for more than sixty years, appeal to me as much as any fancy preserved Pacific. It is not the stuffed *Flying Scotsman* that I chose to illustrate, but the restored T9 4–4–0 No. 120. The 'Greyhounds' as the T9s were known had a life-span of more than sixty years. Although a grey day at Eastleigh is not the most glamorous setting, my conscience allows me to include an engine now in a museum, as she may never again be seen on the track, let alone at a steam Motive Power Depot. The T9s were indeed special engines, certainly for those who love to recall the LSWR in Cornwall.

PLATE 34. 'Schools' 4–4–0 No. 30934 *St Lawrence* was amongst the last batch withdrawn to make this class extinct on 29 December 1962. It had also performed the last SR Western Section main line passenger working, when it took over the 15.20 Waterloo–Weymouth from Basingstoke on 1 November. On 18 May 1963, I saw her in steam at Basingstoke, whence she ran down to Eastleigh, where she is seen here; perhaps the last photograph ever taken of an unrestored Schools in steam on BR metals. *(Following pages.)*

30934

30934
SPL
23

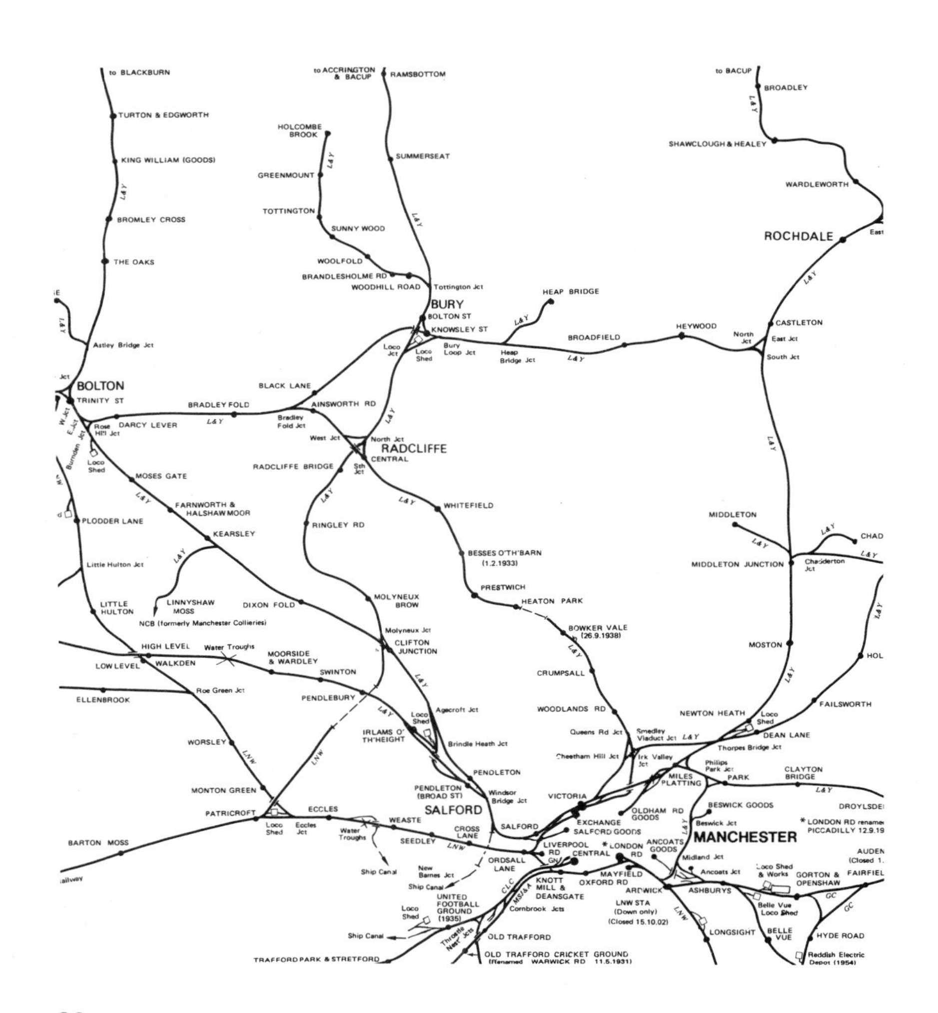

to BLACKBURN
TURTON & EDGWORTH
KING WILLIAM (GOODS)
BROMLEY CROSS
THE OAKS
Astley Bridge Jct
BOLTON
TRINITY ST
W. Jct
E. Jct
Burnden Jct
Rose Hill Jct
Loco Shed
DARCY LEVER
BRADLEY FOLD
Bradley Fold Jct
BLACK LANE
AINSWORTH RD
MOSES GATE
FARNWORTH & HALSHAW MOOR
KEARSLEY
PLODDER LANE
Little Hulton Jct
LITTLE HULTON
LINNYSHAW MOSS
NCB (formerly Manchester Collieries)
DIXON FOLD
HIGH LEVEL
LOW LEVEL
WALKDEN
Water Troughs
MOORSIDE & WARDLEY
SWINTON
ELLENBROOK
Roe Green Jct
PENDLEBURY
WORSLEY
MONTON GREEN
PATRICROFT
Loco Shed
Eccles Jct
ECCLES
BARTON MOSS
to ACCRINGTON & BACUP
RAMSBOTTOM
SUMMERSEAT
HOLCOMBE BROOK
GREENMOUNT
TOTTINGTON
SUNNY WOOD
WOOLFOLD
BRANDLESHOLME RD
WOODHILL ROAD
Tottington Jct
BURY
BOLTON ST
KNOWSLEY ST
Loco Jct
Loco Shed
Bury Loop Jct
West Jct
North Jct
RADCLIFFE
CENTRAL
RADCLIFFE BRIDGE
Sth Jct
RINGLEY RD
WHITEFIELD
BESSES O'TH'BARN (1.2.1933)
PRESTWICH
HEATON PARK
BOWKER VALE (26.9.1938)
CRUMPSALL
WOODLANDS RD
MOLYNEUX BROW
Molyneux Jct
CLIFTON JUNCTION
Agecroft Jct
Loco Shed
IRLAMS O' TH'HEIGHT
Brindle Heath Jct
PENDLETON
PENDLETON (BROAD ST)
Windsor Bridge Jct
SALFORD
WEASTE
Water Troughs
SEEDLEY
CROSS LANE
Ship Canal
New Barnes Jct
Ship Canal
UNITED FOOTBALL GROUND (1935)
Loco Shed
Ship Canal
Throstle Nest Jcts
TRAFFORD PARK & STRETFORD
OLD TRAFFORD
OLD TRAFFORD CRICKET GROUND
(Renamed WARWICK RD 11.5.1931)
HEAP BRIDGE
Heap Bridge Jct
BROADFIELD
HEYWOOD
North Jct
East Jct
South Jct
CASTLETON
to BACUP
BROADLEY
SHAWCLOUGH & HEALEY
WARDLEWORTH
ROCHDALE
East
MIDDLETON
MIDDLETON JUNCTION
Chadderton Jct
CHAD
MOSTON
HOL
FAILSWORTH
NEWTON HEATH
Loco Shed
DEAN LANE
Thorpes Bridge Jct
Queens Rd Jct
Smedley Viaduct Jct
Cheetham Hill Jct
Irk Valley Jct
Philips Park Jct
MILES PLATTING
PARK
CLAYTON BRIDGE
VICTORIA
EXCHANGE
SALFORD GOODS
OLDHAM RD GOODS
BESWICK GOODS
Beswick Jct
MANCHESTER
DROYLSDE
* LONDON RD renamed PICCADILLY 12.9.19
LIVERPOOL RD
GN
CENTRAL
* LONDON RD
ANCOATS GOODS
Midland Jct
AUDEN (Closed 1.
ORDSALL LANE
KNOTT MILL & DEANSGATE
MAYFIELD
OXFORD RD
ARDWICK
Ancoats Jct
Loco Shed & Works
GORTON & OPENSHAW
FAIRFIEL
ASHBURYS
Cornbrook Jcts
LNW STA (Down only) (Closed 15.10.02)
Belle Vue Loco Shed
LONGSIGHT
BELLE VUE
HYDE ROAD
Reddish Electric Depot (1954)
L&Y
LNW
CLC
MSJ&A
GC

8 Clifton Junction

12 March 1849 probably has not impinged itself indelibly on your mind. As a great historic date it comes well down the list as far as most schoolboys are concerned. Yet in its way it typifies the pride, passion, entrepreneurial zeal and dedication that separated the men of the 'Railway Mania' from ordinary mortals. It was the date of the Battle of Clifton Junction.

It is more than 130 years since there was any evidence of battle at 793028, the OS map reference to Clifton Junction on the Seventh Series, One Inch Map Number 101, Manchester. A glance at the 'Liverpool and Manchester' page of the Ian Allan *Pre-Grouping Atlas and Gazetteer* shows where the Manchester-Bolton line, opened by the Manchester and Bolton Railway on 29 May 1838; and the later line to Bury, opened by the Manchester Bury and Rossendale Railway on 28 September 1846, diverge at Clifton Junction. The Manchester & Bolton Railway was taken into the Lancashire & Yorkshire; and the Manchester Bury and Rossendale Railway was incorporated in turn within the East Lancashire Railway. Thus the great contemporary Lancastrian rivals met at Clifton Junction. From Clifton Junction eastwards, for the 5 miles into Manchester, the Lancashire & Yorkshire shared its running powers into the city with the East Lancashire.

'Shared its running powers' sounds a simple, civilised activity in the 1980s. The East Lancashire paid tolls to the Lancashire & Yorkshire for the privilege of through-running, and a scale of charges was agreed between the rival companies. The level of tolls was arranged on the spot, depending on the number of passengers on the East Lancashire trains. It was not long before discussion turned through disagreement, argument and hostility, to a total breakdown of normal commercial arrangements. The Lancashire & Yorkshire then insisted that all East Lancashire trains stop at Clifton Junction for the number of passengers to be checked. The East Lancashire refused, determined to exercise their rights.

Fig. VIII. The area surrounding Clifton Junction.

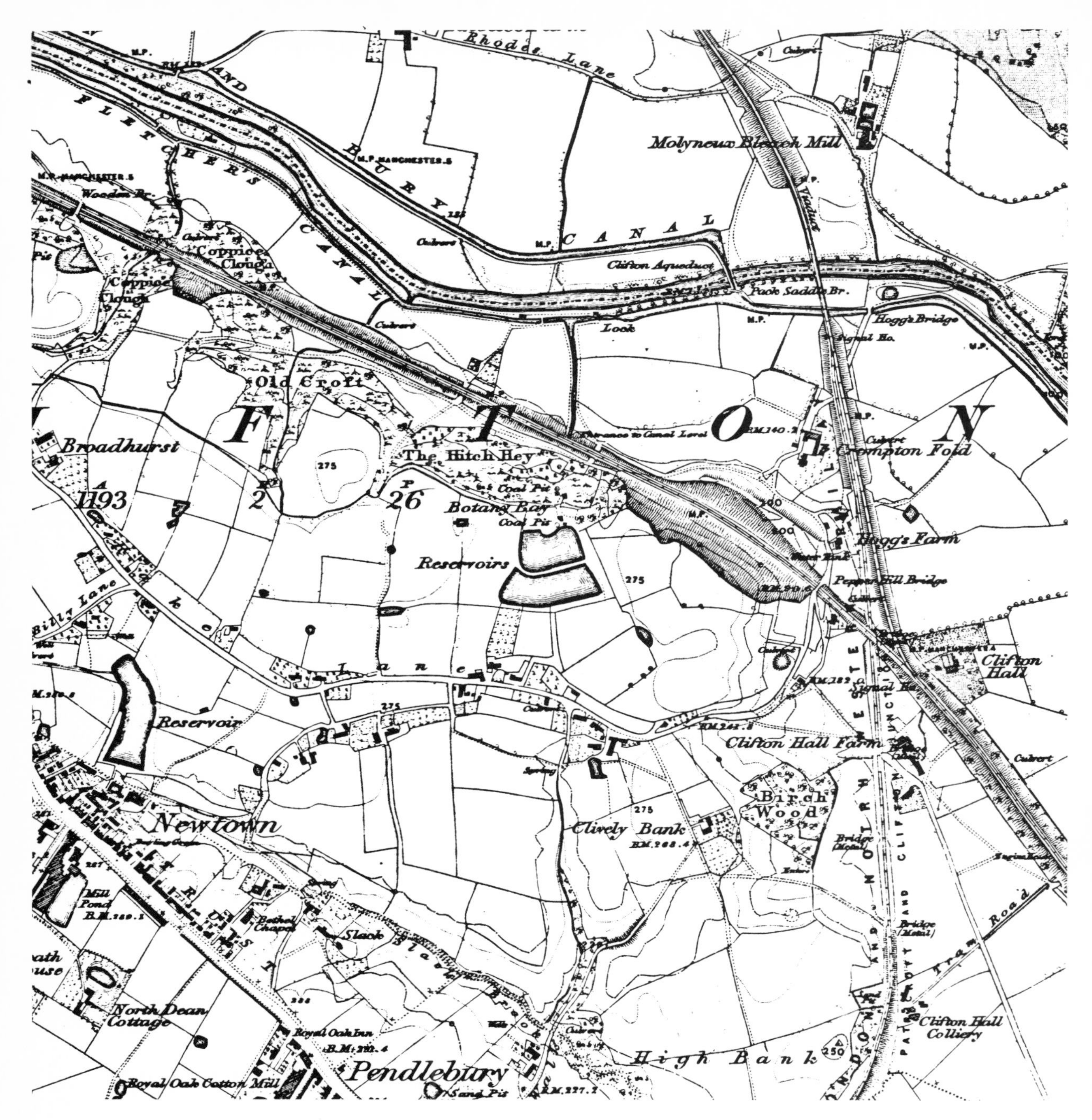

The scene was set for the confrontation that took place on Monday 12 March. The Lancashire & Yorkshire decided to enforce its right to stop all East Lancashire trains, and the track was blockaded ahead of an East Lancashire train due to pass Clifton Junction at 10.30 a.m. The train was forced to halt and Lancashire & Yorkshire staff boarded it to inspect passengers' tickets and enforce the toll. However the East Lancashire, anticipating the move, had already collected the tickets at

Fig. 9 (page 90), Fig. 10 (page 91), Fig. 11 and Fig. 12 (page 92, top and bottom).
With the help of the OS it is possible to trace the development of Clifton Junction from its early rural beginnings to its latter-day urban demise. On the 1850 map the junction itself can be seen sandwiched between Hoggs Farm and Clifton Hall Farm, and adjoining the grounds of Clifton Hall itself. By 1909 Clifton Junction was on the map. The Lancashire & Yorkshire Railway is prominent; and note an interesting detail – the 'Signal House' at Clifton Junction itself, had become an 'SB' or signal box. Nevertheless, the area remains primarily rural. By 1967 Clifton Junction is part of suburban Manchester. It is not the fault of the map-maker that the area has lost much of its cartographic appeal. Finally we see the scene today. The words 'Dismantled Railway' tell their own tale. Clifton Junction is a junction no more, and the M62 motorway cuts a swathe across the scene.

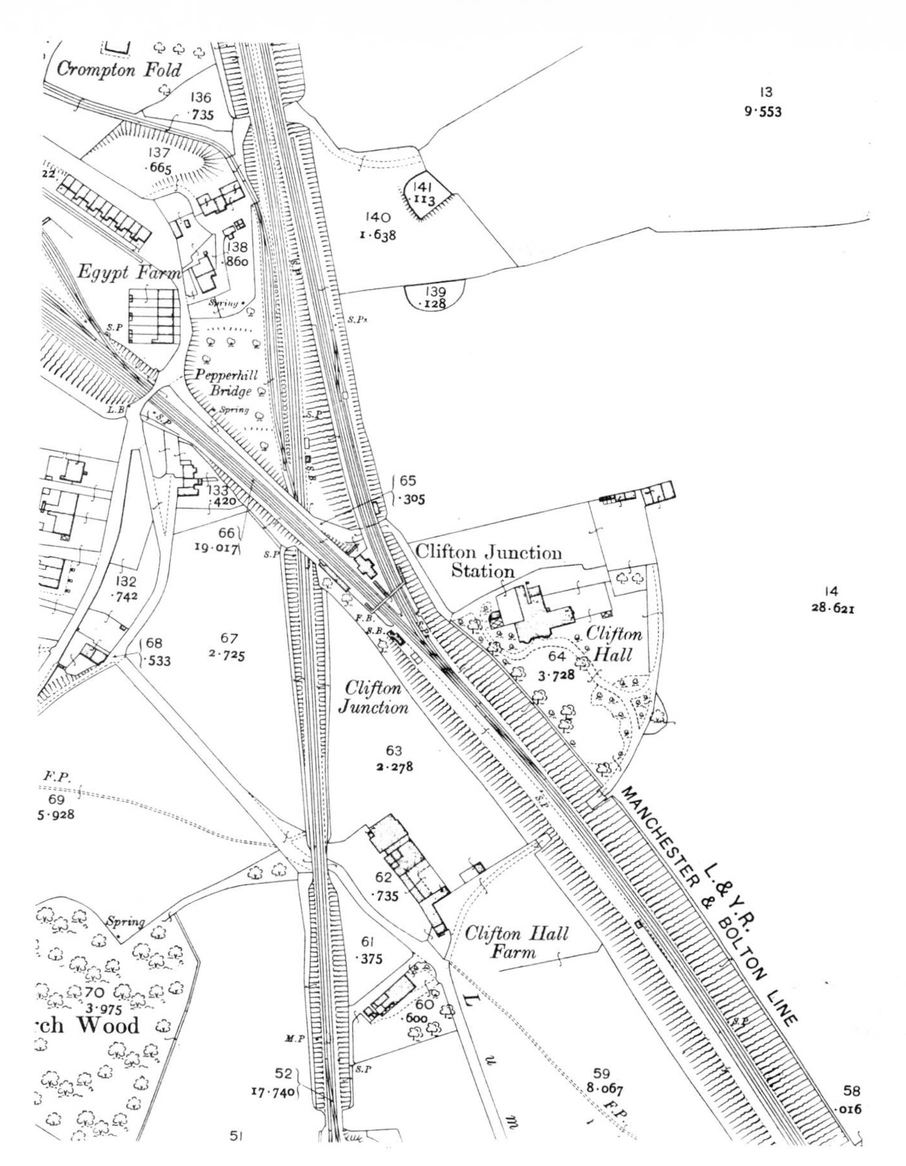

Bingley Road station, north of Clifton Junction on the Bury line. Round one ended thus, all square!

Next, the East Lancashire men set about removing the obstruction: whilst the Lancashire & Yorkshire staff screwed down all the brakes on the East Lancashire train. Round two – deadlock all round.

Finally the East Lancashire men decided to blockade the down line, which was the Lancashire & Yorkshire main line through Clifton Junction to Bolton and Preston. This was the last straw. Nine or ten trains were held up, passengers were infuriated and legal action was threatened. The Lancashire & Yorkshire were suffering more than the East Lancashire so soon after mid-day the Lancashire & Yorkshire relented.

Questions were asked in Parliament, but the President of the

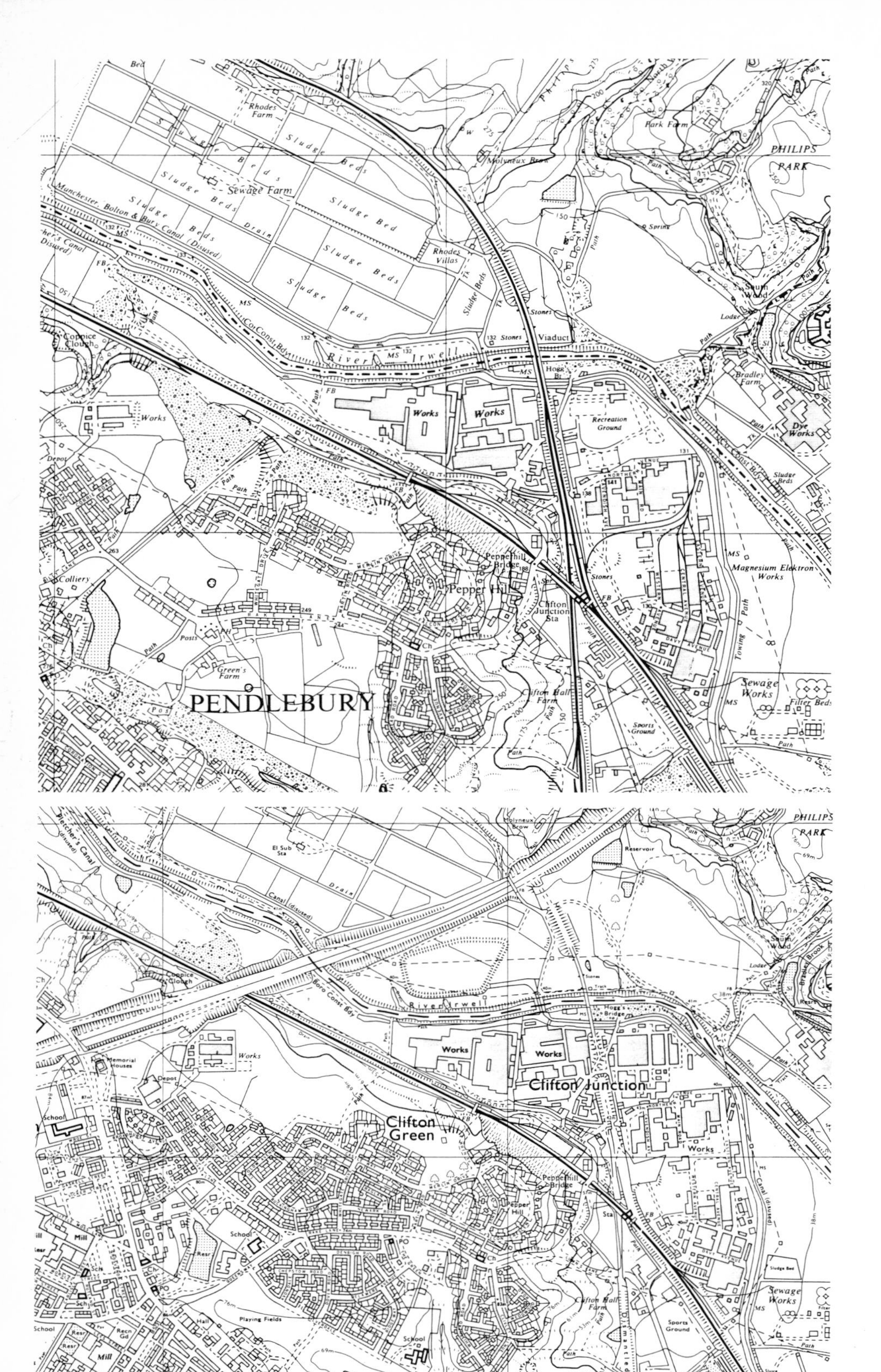

Map captions on previous page.

PLATE 35. A classic L&Y run: Class 5 4–6–0 No. 45096 hurries commuters from Manchester through Clifton Junction with the 17.45 to Wigan Wallgate, in April 1964. At this date, the Junction still saw a stream of steam-hauled commuter trains in the rush-hour on the main line westwards to Bolton, Crow Nest Junction, Wigan, Southport and Liverpool Exchange, but a look at the contemporary working timetable shows that almost all passenger services taking the old East Lancashire line north at the Junction, to destinations like Colne, Skipton and Accrington were already relegated to d.m.u. haulage. *(Opposite.)*

PLATE 36. Built at Derby in 1938, Stanier 2–6–4T No. 42626 passes Pepper Hill Signal Box, west of Clifton Junction, with the 17.47 Horwich to Manchester Victoria train on 22 April 1964. Many of the branch and secondary services in Lancashire were handled by these handsome and fast tank engines. In the hands of the right driver, they could out-accelerate and outpace the d.m.u.s that replaced them. Pepper Hill had carriage sidings as well as a signal box. The station at Clifton Junction handled goods as well as passenger and parcels traffic, live-stock, horse-boxes and Prize cattle vans. It had a fixed six-ton crane. *(Following pages.)*

PLATE 37. Good fortune, not immodesty leads me to claim this as the first and only published colour photograph of a Stanier 2–6–0 hauling a passenger train! Known variously as Lobsters or Camels, these elusive engines – only forty were built – were Stanier's second new LMS design, introduced in 1933. Here, No. 42947, one of the original ten engines built with safety-valves located on the boiler, hurries through Clifton Junction with the 17.15 Manchester Victoria to Blackpool Central train on 22 April 1964. With low horizontal cylinders and shallow section running-plate they were not Stanier's most attractive design.

Board of Trade parried questioning on the grounds that public safety, not convenience, was his responsibility, and the former had not been endangered.

Relations between the Lancashire & Yorkshire and the East Lancashire remained bad. In fact they deteriorated to such an extent by 1852 that the East Lancashire drew up plans to build their own line from Clifton Junction to Manchester, to run parallel to the Lancashire & Yorkshire line as far as Salford. This finally ensured that sanity should prevail, and the Lancashire & Yorkshire agreed that the Clifton – Salford section should become joint property. Powers to bring this about were sanctioned on 3 July 1854.

How dull it all seems today. Yet without the in-fighting of the railway mania, we would have an even smaller railway system than that which we have today. The East Lancashire Railway has long since faded from living memory. The Lancashire & Yorkshire merged with the LNWR in 1922, and both were enveloped within the LMS at the 1923 grouping. In turn the LMS was swallowed up by British Rail in 1948. Compare the gazetteer to which I referred earlier, to the Oxford Publishing Company's current *Rail Atlas of Britain*, and it is evident how decimated is the Lancashire railway landscape.

PLATE 38. An early 'Black 5', No. 45196, built with domeless boiler by Armstrong Whitworth in 1935, leaves Clifton Junction for Manchester Victoria with the 16.36 from Preston on 22 April 1964. Arguably the most versatile engine ever built, Stanier's classic mixed-traffic engines outlasted every other class in service on British Rail, alongside his '8F' freight locomotives.

PLATE 39. A steam-hauled freight leaves a trail of smoke hanging on the still evening air as it drifts across Molyneaux Brow viaduct, on the ELR line in April 1964. Soon, steam will disappear: the viaduct will carry trains no more and the word 'Junction' will no longer be affixed to 'Clifton' in BR timetables. In steam days, the 17.40 (Sx) Victoria to Bolton Trinity Street took sixteen minutes; to-day, the 17.45 (Sx) d.m.u. takes nineteen minutes. In steam days, the 18.10 (Sx) Victoria to Wigan Wallgate arrived at 18.37; to-day the 18.10 (Sx) Victoria to Wigan Wallgate arrives at 18.45. Anyone for steam . . . ?

The line from Clifton Junction to Bury has gone and Clifton is a Junction no more. The sight of a long freight-train drifting across Molyneux Brow viaduct on a summer's evening, leaving a trail of smoke in the still air, is but a memory. Gone too are the Crabs, with their Lancashire & Yorkshire chimneys that even I can remember, rushing West from Manchester Victoria on trains such as the 17.40 to Blackburn. Gone too are the Black 5s, the Stanier 2–6–4Ts, the Austerities.

'Bolton to Bury by train, sir? It must be some long time since you were here . . .' Who can remember the once-famous 4.25 pm express (not 16.25) from Salford to Colne? Never again shall we see those splendid Aspinall/Hoy/Hughes Lancashire & Yorkshire 2–4–2 tank engines slow to 30 mph at Clifton Junction, then cross the viaduct and charge the 1 in 96 bank up to Bingley Road and on up the fearsome ascent from Bury to Baxenden. . . .

9 SECR and the Maunsell Moguls

By the beginning of 1963, rail enthusiasts searching for steam were finding it increasingly difficult to be selective about location. Ideally one sought lines on which elderly engines ran a regular and reasonably frequent steam-hauled passenger and freight service through congenial countryside, preferably with access to Motive Power Depots. For spice, it helped to have the trains interspersed with the occasional 'foreign' engine (see Chapter 12). With British Rail's regions vying to be the first to eliminate steam, and with main line passenger services being the earliest targets for 'dieselisation' the secondary and cross-country services were an obvious target for contemporary railway photographers. The South Eastern and Chatham Railway (SECR) line between Tonbridge and Reading (Southern), via Redhill and Guildford, fulfilled most of the necessary criteria.

The line has an interesting history. In the fiercely competitive era in which it was constructed the SECR poked its finger a long way from its Kentish heartland to reach Reading. Its junction there with the GWR created an interesting through-route from the North-West to the South-East. Trains from Birkenhead to Hastings, or Wolverhampton to Margate were living testimony to the imaginative marketing skills of the independent pre-grouping railway companies. As early as 1846 the Gauge Commissioners (a sort of early Victorian Railway National Enterprise Board!) recommended the creation of a rail link from Oxford to Reading and Basingstoke in order to provide a through-route from the North to the South-West. By the end of 1856 a standard-gauge line had been built enabling trains to run via the GWR from Birmingham through to Southampton and Portsmouth. The success of the 'Reading Connection' with the South Western prompted the construction of a connection at Reading with the South Eastern Railway (SER). On 1 July 1863 the connection with the SER was used for the first time by a regular passenger service. It was one of the first cross-country

PLATE 40. Many of the smaller stations retained their rural atmosphere, belying the stockbroker-belt development surrounding the communities that bear their names. Gomshall and Shere was a case in point, and this picture shows a familiarity and rapport between engine crew and general population that can never be replaced. BR/STD Class '4MT' 2–6–4T No. 80137 awaits the 'right away' with a Reading (Southern) to Redhill train, on a warm afternoon in August 1963.

inter-company routes in Britain. The south-bound train left Birkenhead at 08.05, running via Birmingham, Oxford, Reading and Redhill, reaching Hastings at 18.30 and Dover at 18.50.

Various interesting permutations of service continued for almost exactly a century. Over the years, through the operation of slip coaches and modernisation and extension of track layout at Reading, the Birkenhead to Dover train survived, with portions for Ramsgate, Margate, Deal, Hastings, Brighton and other destinations. In the 1930s the train departed Birkenhead Woodside (see Chapter 13) at 08.00 and ran to Sandwich, arriving 16.42. On 5 September 1964 the last through-service was run, 101 years and 2 months after its inception (see Plate 22). For the last thirty-five years of the service, the Maunsell Moguls played their part in providing motive power.

The line just survived the Beeching axe; a decision for which we shall have cause to be grateful in years to come, as I am certain that the Channel Tunnel will eventually be built and that the SECR line will then enter an era perhaps dreamt of by its progenitors. As a 'London Rail By-Pass' in future years, the line's prospects seem to be bright indeed. It is no coincidence that the long-delayed M25 motorway follows a not dissimilar course for much of its actual and proposed length.

Let me return to the 1960s. My first attempts to locate steam in and around London, at the start of my 'dream of steam' at the very end of 1962, led me firstly to the London termini. Talking to a railwayman at Victoria, he told me to 'try my luck' on the SECR line, the existence of which I was barely aware. My memory, prodded into action, reminded me of the sight of wisps of steam in the Redhill area, flashing by on the Quarry line, when I was commuting between Brighton and Victoria in the 1950s. Occasionally, sitting in a corner seat on a '4–LAV' on a Brighton semi-fast, I had noticed an engine in the immediate vicinity of Redhill station. So I set out to investigate.

Professional railway photographers and experienced enthusiasts may well scoff at my ignorance. My only excuse is that the long dormancy of my enthusiasm had eroded my contemporary knowledge of railway matters. Standing on the down platform at Redhill, on the dull morning of 15 April 1963, I was delighted to observe the occasional 'steamer' amongst the electrics. Enquiry quickly told me of another world beyond Reigate and the Redhill/Reigate shuttle operated by 2–BILs.

I took a few shots from my platform-end location at Redhill that day; but they are unworthy of inclusion in this book. Later, as I left Redhill and returned home to Chiswick, I knew I had to explore the distant lands that lay to the west of Reigate! Armed with OS map No. 170, Seventh Series, London SW, I persuaded Jane that we should picnic in deepest Surrey whilst I took photographs! I shall never forget 20 July 1963. It was not my first glimpse of the line, but it was the first time I had persuaded Jane to join me in devoting a whole day to railway photography, and the first time I was able to plan a day on the line, rather than snatch the odd half-hour whilst my impatient spouse champed at the bit. (My first few shots on the SECR, taken some few weeks previously, were taken near Edenbridge, before I even became aware of the history and significance of the line.)

My pulse quickened as, armed with map and current timetable, we picked a point that seemed to have the right ingredients. Parking the car in the lane at map refernce 113483, and with my 'Spotters Bible' (*abc British Railways Locomotives* to give it the dignity it deserves), we scrambled down the cutting and on to the track.

I never imagined, as mid-day approached and I suddenly glimpsed a puff of smoke followed by the unmistakeable beat of a steam engine, that I should be writing a chapter in a book some seventeen or eighteeen years later, about that moment. My memory sometimes lets me down, but my trusty 'catalogue' faithfully records the details of the train that approached . Entries J30/31 indicate that I hastily shot two pictures of the train. As it

31816

PLATE 41. Guildford remains a busy cross-country as well as a commuter station. Whilst Ivatt tanks no longer simmer quietly on LB&SC Horsham line trains, Western Region d.m.u.s at least add some variety to an unremitting stream of electric trains. Steam remained on the SECR line through Guildford into 1965 – just. Maunsell 'N' mogul No. 31816 approaches the station on 2 January 1965, the last weekend of steam operation, with the 12.32 Redhill to Reading (Southern), passing the engine shed, the site of which is now a car park.

passed, I scribbled the number down on the slightly scruffy piece of paper on which I always recorded daily events I sought to memorise.

'31871', said Jane, 'what is it?'

'I'll have a look'.

I checked it and told her it was an 'N'.

I have in front of me as I write those words, the number book which told me it was a 4P5F N&NI 2–6–0 N, introduced 1917. Maunsell SEC mixed traffic design. I had photographed my first Maunsell Mogul-hauled passenger train. It was an agreeable experience, often to be repeated over the next few years. I took every opportunity to spend time on the line, taking Jane with me whenever I could.

The Maunsell Moguls were splendid engines. It took me some time to learn to differentiate between the 'N's and the 'U's. I never saw an 'N1' as all six were withdrawn in November 1962, and my very first Maunsell mogul photograph, according to my catalogue, was a shot of 31866 near Edenbridge hauling a short freight in March 1963. On that occasion, Jane and I had been *en route* from our Chiswick flat to the Old Surrey & Burstow point-to-point and I was 'allowed' ten minutes 'by the track'. On another fleeting visit with Jane, *en route* to friends near Edenbridge, I found an eerie and remote place, at which I took a strange picture. On the Kent-Surrey border, a mile to the west of Edenbridge, the SECR, in a cutting, crosses the former LBSCR line from Eridge and Tunbridge Wells to Oxted. The LBSCR is in a tunnel, from which it emerges momentarily to pass under the SECR line. It is a strange, isolated place not often visited. The light for colour photography is generally poor, but on 23 March 1963 I captured in my lens a Wainwright SECR 'H' class 0–4–4 T emerging fleetingly from the tunnel with a Tunbridge Wells to Oxted train. This place is shown by Derek Cross in 'Double-Headed Trains-South': I am pleased enough to sit at the feet of the Master. . . .

The Maunsell Moguls provide me with fond memories, for we spent many happy hours together. They were friendly engines. Shedded along the length of the SECR line, at Tonbridge, Redhill, Guildford and Reading Southern, I was to capture many of both classes 'N' and 'U' in the next two and a half years. I remember seeing them on shed at Nine Elms, Feltham, Basingstoke, Salisbury, Yeovil, Brighton and Stewarts Lane as well as at the four sheds mentioned above.

The history of the Woolworths, the Ns built at Woolwich Arsenal after World War 1, is told briefly in Chapter 4. The fact they were built was a commentary on the success of the first 'N', which appeared from Ashford, main works of the SECR, in 1917.

PLATE 42. The association of ideas generated by the juxtaposition of the word 'Reading' with the phrase 'motive power depot' meant, to most people, the GWR shed. The Southern shed, opened when the South Eastern Railway took over the Reading, Guildford and Reigate Company, lasted until the end of steam on the SECR in January 1965. By then it was merely a stabling-point and sub-shed to Guildford. Less than two months life remain, as the setting sun casts a wintry light on Maunsell 'U' class 2–6–0 No. 31790 on 7 November 1964.

Maunsell became CME of the SECR in 1913, and one of his first acts was to recruit two of the current 'young Turks' from the design department at Swindon. To these two young designers, H Holcroft and G H Pearson may be attributed the seeds of the moguls, as only the GWR was at that date using the 2–6–0 wheel-arrangement for its 'intentional' mixed-traffic engines; plus, of course, the taper boiler that was to become such a feature of successful British locomotive design.

The Ns lasted almost to the end of steam on the Southern, No. 31405 not being withdrawn until June 1966. Eighty engines were built between 1917 and 1934. The original engine served SECR, Southern Railway and Southern Region of British Rail for nearly forty-seven years, and it was the demise of steam itself, rather than of the Ns, that saw their end. Although introduced in 1917, it was still a thoroughly modern engine.

The Ns success begat the ill-fated 'River' class 2–6–4 tanks, whose life was short-lived following the Sevenoaks disaster of 24 August 1927, when engine number 800 *River Cray* left the track at speed between Dunton Green and Sevenoaks. Smashing into a bridge, there were many killed. Subsequent tests, on both the LSWR and GN main lines of the SR & LNER, showed no basic design fault, but clearly indicated that the Rivers were extraordinarily dependent on sound track for safe running. Water surging in the tanks probably caused the problem. The Glasgow and South Western Baltic tanks suffered from the same fault, and their 'rolling' attributes caused their restriction to the main Glasgow to Ayr line with its good track.

The Sevenoaks disaster gave the Rivers a bad name with the engine crews and with the public. Maunsell decided to cut his losses and the Rivers were withdrawn and totally rebuilt, from 2–6–4 tank to 2–6–0 tender engines. Classified as the new Class 'U' they formed the basis for another good reliable mogul. In 1928, the twenty Rivers, now reincarnated as Us, emerged, some from Ashford, some from Brighton and some from Eastleigh. None of the rebuilt engines were withdrawn until June 1963, when 31794 and 31795 went. No. 31791 formerly No. A 791 *River Adur* lasted until June 1966, becoming, together with 31639, the last Us in service. The latter engine was one of thirty Us built by the Southern Railway at Brighton and Ashford between August 1928 and May 1931. To the older men, the Us remained known as Rivers, and are sometimes still recalled as such.

Between June 1928 and November 1931, Maunsell had built twenty-one three-cylinder variants of the Us – the U1s. They were less good looking than the Us, however, having a flat front in order to accommodate the third cylinder.

To complete the catalogue of Maunsell's moguls, mention must be made of the three-cylinder 'N1's. The first one appeared immediately prior to the 1923 grouping, the remaining five emerged from Ashford in 1930.

As I mentioned earlier, it took some time to differentiate, without reference to the numbers, between 'N's and 'U's. The 'U's had 6′ 0″ driving wheels, and small splashers, whereas the 'N's had 5′ 6″ driving wheels without splashers. Notwithstanding the availability of more modern motive power, including many British Rail newer 'Standard' classes, the Maunsell moguls continued to work both freight and passenger trains, not only on the SECR but also on the SW main line until 1965. Throughout their life they hauled much freight around the Home Counties. To the end, they remained popular, not only with enginemen for their surefooted reliability, but also with the operating department, both for their reliability, route availability and for their fuel economy and ease of maintenance.

44181

10 The Vanished Shed Scene

PLATE 43. Warrington merits its own book. With two east/west lines, the LNW & CLC, bisecting the North-South trunk-route of the West Coast main line; with Bank Quay recalling its nautical past as a Mersey port; and with the Ship Canal nearby, the railways accentuated its natural importance. It deserves far more attention than it has received in the past. Arpley as well as Dallam MPD was open in the winter of 1962, and as late as 2 December 1964 Dallam still looked, smelled and felt like the incarnation of the steam railway. The discerning eye will pick out the vintage features, but will note the yellow stripe on the cabside of LMS '4F' 0–6–0 No. 44181, an indication that engines thus marked were banned from working 'beneath the wires' on the overhead-electrified lines south of Crewe, a notification heralding the death of steam.

The steam running-shed, or motive power depot (MPD), epitomised the intangible attraction of the steam railway. It was not just the seductive smell of steam, smoke and oil that created the irreplaceable atmosphere. An engine shed was a place where evidence could be found to testify to a conviction held by many, that the steam engine has a soul. It was here that man and machine merged their personalities, by sharing common needs; it was here that steam engines proved they were the nearest man-made invention to resemble a living thing.

In common with breathing creatures, working steam engines had to be looked after. They needed food (fuel) and water, and at the end of the day they returned to their 'home' shed or were billeted elsewhere as 'visitors'. Ablutions and toilet facilities were represented by the need to have tubes and grate cleaned regularly, involving the disposal of ash. Regular lubrication was essential!

Sheds came in two main shapes, the more common being the 'straight through' style, where parallel tracks ran through or terminated in a shed which straddled a number of tracks, from a single 'road' in a tiny rural outpost to major sheds like Nine Elms. The other variety was the roundhouse. Here a central turntable enabled engines entering the shed to be fed into one of the available slots which radiated from the turntable.

Most British roundhouses were fully-roofed (Plate 13), while some, like Guildford, had the turntable in the open air with the roads under cover – more in the continental style (Plate 45). Some of the most important sheds, such as Derby, Cricklewood, Nottingham, Old Oak, Swindon and York, had more than one roundhouse, with connecting access one to the other. The atmosphere in a large roundhouse was akin to that of a cathedral: lofty, calm and contemplative. Indeed the murder of steam was akin to the desecration of the monasteries – at least in architectural terms.

PLATE 44. Friar Tuck lives! Bright sunshine and a cheerful face hide the imminent closure of yet another Merseyside shed, Speke Junction. Designated 8C, the shed was in the triangle of the Allerton/Garston Docks/Ditton Junction lines. On 10 August 1967 one of the depot's 9F 2–10–0s, No. 92054, returned to base from one of the shed's freight rosters. This engine was one of the last 9Fs to remain in service, being withdrawn in May 1968.

The big engine sheds, particularly the Victorian roundhouses, were architecturally important. Thus the London and Birmingham Railway's Camden roundhouse is a scheduled building and has been externally preserved. The roundhouses, like the great railway stations, were built by men imbued with a spirit of imperial pride. Sadly, whilst many stations still survive, most of the great roundhouses have gone, torn down, in a fit of economic vandalism, by men also motivated by the determination to eradicate every vestige of the steam era. These were little men, whose vision extended the length of a slide rule.

If the men who destroyed the architectural evidence of the steam era were small in spirit, the men who worked inside those buildings were big: men of determination, men with a fierce pride and loyalty to 'The Company'. Seldom in mankind's span has there been a relationship to compare with this kinship. If you can, read what Brian Haresnape has to say in *A Journey by Design,* referring to the relationship between man and the steam-engine (p 79 para 2).

Every engine shed had one thing in common by the end of steam – intolerable working conditions for the long-suffering staff. Crawling in and out of the smoke-box of an engine was not an idyllic occupation even when the engine was in its prime;

PLATE 45. Guildford shed was still a busy place on 25 May 1963, when 'USA' 0–6–0T No. 30072 was shed pilot. The shed was unusual as its central open-air turntable gave access to a covered roundhouse. This engine is now preserved and working on the Keighley & Worth Valley Railway, and looks slightly weird in its brown livery. *(Top.)*

PLATE 46. 15 May 1963 was a splendid day, as shown by this picture of rebuilt 'West Country' 4–6–2 No. 34013 *Okehampton* on shed at Brighton. She was a particularly fine specimen, and was often present at historic events. She headed the last steam journey down the Uckfield line from Victoria to Brighton, the last steam 'City express' from Cannon Street to Ramsgate on 12 June 1959, jointly headed the last farewell tour of the Somerset & Dorset on 6 March 1966, and headed the last steam 17.30 Weymouth–Waterloo on 7 July 1967. 34013 was called 'the fireman's friend' at Brighton MPD, where she was shedded at this time. She was transferred to Salisbury MPD when Brighton lost its Pacifics in September 1963.

PLATE 47. My apologies to LNER enthusiasts for the lack of their favoured engines in this book. I hope I am not adding insult to injury by featuring Thompson 'B1' 4–6–0 No. 61237 *Geoffrey H Kitson* at Midland Railway HQ on 9 September 1966. A Wakefield engine, she was one of the few B1s named after men connected with the constituent companies; perhaps her visit to Derby rekindled the ghosts of the old Great Northern! She was withdrawn two months later. *(Following pages.)*

when the engine was on its last legs and funds for cleaning were but a memory, then only one factor kept these men at work – dedication to the steam engine.

Steam traction was dirty at the best of times; and 1962–1968 were the worst of times. Yet, surrounded by ash, dirt, soot, oil, coal and water, and working with obsolete equipment, the spirit of the staff never seemed to fail. One day, before it is too late, a proper tribute should be paid to the last men of steam. They remain a breed apart.

61237

34076
34076

11 Bulleid's Last Fling

PLATE 48. An unrebuilt Bulleid 'Battle of Britain' Pacific No. 34076, *41 Squadron*, enjoys the sunshine at Feltham. Argument still rages over the merits of the original Bulleid designs and BR rebuilds, described by Derek Cross as 'bastardised versions which were poor lame shadows of a magnificent concept'. Bulleid was a latter-day Brunel, but he lacked the stage supplied by the great entrepreneurs of the Victorian era. *(Opposite.)*

PLATE 49. Third-rail electrification has reduced Winchester to a commuter outpost. The interest that existed there in the steam era has long since gone. On 14 July 1965, the sun glinted on 'West Country' Pacific No. 34100 *Appledore*, heading a Waterloo to Bournemouth Express. She was based at Salisbury, having been rebuilt by BR at Eastleigh in September 1960. *Appledore* survived to the end of Southern steam, 9 July 1967, when she was stored at Nine Elms pending disposal, having completed 712,916 miles since entering service in December 1949. She was sold to Cashmores, Newport, and by September 1968 had been broken up. *(Following page)*

Nobody who observed the circumstances in which steam's last main-line duties were performed in 1967, on the Waterloo/Southampton/Bournemouth/Weymouth services can have failed to be impressed by two factors; firstly, the dedicated determination with which the footplatemen tackled their melancholy task; and secondly, the manner in which ill-kempt, neglected engines responded to their final challenge.

Immediately prior to the introduction of the electrified service on the Waterloo–Bournemouth trains in July 1967, the line saw an epidemic of speed-restrictions, diversions from fast to slow lines and other circumstances which militated against efficient railway operation. Add to this, the almost total ban on expenditure on maintenance of the remaining steam stock before withdrawal and fast running should have been an impossibility. Yet it happened daily. The Nine Elms, Eastleigh, and Bournemouth men willed their steeds to heroic efforts, and Bulleid's fine Pacifics – Merchant Navies, West Countries and Battle of Britains – responded above and beyond the call of duty. The July 1967 *Railway Observer* reports on page 234, '*The general level of steam running between Southampton and Waterloo has perhaps never been higher than in April and May of this year*'.

On 19 May, No. 34104 *Bere Alston* with eleven coaches, 395 tons gross, came from Basingstoke to Woking start-to-stop in exactly 21 minutes for the 23½ miles, touching 85 mph at Fleet and 86 at Brookwood. On 28 April, working the 15.30 Waterloo to Weymouth, No. 35013 *Blue Funnel* left London 14 minutes late with twelve coaches and a van, and arrived in Winchester one minute early, having covered the 66½ miles in 65 minutes.

Earlier, in February, No. 34089 *602 Squadron* with nine coaches loaded to 325 tons gross was badly delayed from Eastleigh to Worting Junction but was then able to accelerate, reaching 92 mph at Fleet and producing a net time of 72¾ minutes from Southampton to Waterloo.

34100
392
392

PLATE 50. Vandalism combined with official neglect created an increasingly difficult task for footplatemen and shed-staff alike, who wanted to end the steam era with spirit. Neglect has been overcome as No. 34089, *602 Squadron*, gathers speed as she passes Nine Elms with the down *Royal Wessex*, on 23 March 1967. *(Above.)*

'Merchant Navy' No. 35023, *Holland-Afrika Line*, which had 'done the ton' on a rail tour to Salisbury on 15 October 1966, reached 92 mph with a heavy load at Porton. In truth, running at 80–90 mph was commonplace at the time, and recording too many examples may make this frequent occurrence sound exceptional, so I shall desist. Steam played a noble part to the end.

The final day of regular express passenger steam on one of Britain's main railway lines was 9 July 1967. To No. 35023 *Holland-Afrika Line* fell the sombre honour of powering the last steam working from Waterloo, prior to complete electrification and dieselisation. She was chosen because she had acquired a reputation for very fast running in the final weeks. Like a great sportswoman, she was in form. As she sped west she bore the legend THE END-THE LAST ONE on her smokebox door. The train was crammed with enthusiasts whose feelings do not need to be described by any words of mine.

Together with the few remaining Merchant Navies, Nos. 35003/7/8/30, No. 35023 was withdrawn that day for scrap. Nine Elms (70A), Guildford (70C), Salisbury (70E) and Weymouth (70G) were closed completely as from midnight on 9 July. From the same time, Feltham (70B), Eastleigh (70D) and Bournemouth (70F) were closed to steam. An era had ended, but to the end, in

PLATE 51. The first of the Merchant Navies, albeit rebuilt, No. 35001 *Channel Packet* hurls itself westwards with the 15.00 Waterloo to Plymouth and Ilfracombe, near Wilton Junction on 25 August 1963. Reminiscent of the great steam racing machines that carved a swathe across Wessex through the years, and like a massive volatile bonfire on wheels, Bulleid's great-hearted engine is a fine memorial to the last Pacifics designed prior to nationalisation.

PLATE 52. Glory of Steam on the Southern. A Bullied Pacific in full flight with a heavy load was an impressive sight. On 25 May 1963 the Atlantic Coast Express was still a recognisable descendant of its great forerunners. No. 35028 *Clan Line* now preserved and restored to BR main line running, is here seen gathering speed as she passes Loco Junction, Nine Elms. *(Opposite.)*

spite of all the difficulties, the best traditions of steam had been upheld. Bulleid deserved no less.

The last 'Merchant Navy' officially to be withdrawn that day, was No. 35030 *Elder Dempster Lines* from Nine Elms; the last 'West Country' was 34095 *Brentor* from Eastleigh; the last 'Battle of Britain' was 34090 *Sir Eustace Missenden, Southern Railway* also from Eastleigh. That same depressing day saw the extinction of other classes too: less distinctive than the Bulleid Pacifics, but still actors at the last curtain-call. No. 77014 (70C) was the final BR Standard Class '3' 2–6–0; No. 80152 (70D) was the last BR Standard Class '4' 2–6–4T; No. 41320 (70F) was the last Class '2' 2–6–2T; No. 82029 (70A) was the last Standard Class '3' 2–6–2T; and No. 30072 (70C) was the last of another distinctive class, the 'USA' 0–6–0T. The latter engine, now preserved on the Keighley and Worth Valley Railway (KWVR) was shed-pilot at Guildford MPD when I photographed her (Plate 45) on 25 May 1963.

I salute them all.

LOCO JUNCTION

393

12 Foreigners

Steam enthusiasts, indeed railway enthusiasts generally, are nationalistic in our passion. Huge American engines, even sleek French models, win few friends and fewer devotees in Britain. Our nationalism tends to the parochial in taste. No Great Western enthusiast will agree that a Stanier Black 5 was the best mixed traffic engine ever built, any more than an LMS man will eulogise a Gresley Pacific in comparison to an LMS 'Princess' or 'Coronation'.

Strange then that most enthusiasts simultaneously disliked foreign steam, passionately espoused their own British favourites, yet were excited by the sight of another company's or region's engine seen on unfamiliar tracks; or indeed by the sight of any operationally – unfamiliar engine. Indeed, a glance at *Railway World* or the *Railway Magazine* will illustrate the point, as correspondents today report 'a "Cardiff" Class 47 seen as far afield as Durham'.

Inter-company trains can be traced back to the Rainhill era, albeit often dogged by jealousy and argument which sometimes reached ridiculous proportions. Since nationalisation in 1948, the idea of competition on Britain's railways has receded into oblivion; indeed the elimination of duplication has been BR policy for many years. But the 1923 grouping, which saw the creation of the 'Big Four' railways in Britain – GWR, LMS, LNER and SR – saw the creation of conditions for national competition within the bounds of safety and sanity.

Over the years, certain places became renowned as changeover points for inter-company or inter-regional trains, Oxford for the GWR/SR link, or Bristol for an LMS/GWR exchange. Yet occasional through workings still brought 'foreign' engines on to unfamiliar metals, often necessitating an extra man in the cab. These visits created interest for most railway enthusiasts, from the humble 'spotter' with his book of numbers, to those who traced and recorded the working life of individual engines. For

PLATE 53. On 25 June 1965, between Fleet and Winchfield, Crewe South's 'Black 5', No. 45046 makes good speed down the SW main line with a semi-fast from Waterloo. This engine first reached the Bournemouth area on a pigeon special from Carlisle, and was temporarily appropriated by the shed at Bournemouth, from where it worked a number of trains between Bournemouth and Waterloo in mid June 1965: a genuine 'foreigner'. *(Opposite.)*

PLATE 54. Christmas Eve 1964, and an unlikely present was the sight of Maunsell 'S15' 4–6–0 No. 30838 alongside Stanier '8F' 2–8–0 No. 48314, on a really murky day at Feltham. *(Following pages.)*

30838

48314

the railway photographer, the opportunity to record a 'foreign' visitor was an exciting event. To have done so in colour makes a picture rare indeed.

Thus, the excitement of capturing in my lens, No. 45046 LMS 'Black 5' on a passenger working on the LSWR/SR main line, on 25 June 1965, was considerable, especially as this was not an inter-regional train, which might have reached SR tracks from the Western Region via Reading and Basingstoke. No. 45046 reached Bournemouth on a pigeon special from Carlisle. It was appropriated by the shed, when on at least two occasions it worked out and back turns from Bournemouth to Waterloo.

Long-distance freight workings sometimes produced the unexpected, and the stamina of the Stanier '8F' 2–8–0 was unsurpassed. These rugged engines, 841 of which were built for service by the LMS and latterly by BR, found their way almost all over the LMS system and far beyond.

The visit of 8F No. 48314 to Feltham on the murky Christmas Eve of 1964 was captured in the lens of my faithful Voigtlander; it was alongside another lesser-known stalwart, and even less glamorous class of freight engine, the 'S15', No. 30838.

In true accord with the time-honoured occupation so beloved of civil servants, the bureaucrats who got their hands on BR after nationalisation seemed to enjoy creating change for its own sake, and inevitable confusion was thus bred. One manifestation of these changes was the transfer of areas from the control of one region to another. Thus Southern Region lost control of their lines to the west of Salisbury to the 'old enemy', the GWR. No sooner did Western Region gain control of their former rival's territory, than they set about imprinting their own image on their new domain. As the steam engine was the ultimate manifestation of railway power and authority, the GWR replaced SR engines with GWR types *quam celerime*.

What the GWR gained in the West, they lost in the North-West. Through services from Paddington to Birkenhead declined, and finally disappeared, as Midland Region took over, and LMS engines reigned supreme in the Chester and Birkenhead areas by the mid 1960s, save for the British Rail Standard classes. By 1965 a copper-capped chimney in Cheshire was almost as rare as snow in the summer, although it would have been a great deal more welcome. How No. 7924, *Thorneycroft Hall*, came to be on shed at Birkenhead on 11 July 1965, I never did discover – but she was a welcome change from Black 5s, 8Fs, 9Fs and Jinties.

No chapter on 'foreigners' would be complete without mention of the Somerset and Dorset. Old hands on the S&D never came to terms with the Western Region takeover, although GWR pannier tanks and Collett 0–6–0s seemed both suitable and satisfactory on

PLATE 55. A foreigner in time alone. GWR engines had long since disappeared from Birkenhead's shed allocation when 'Modified Hall', No. 7924 *Thornycroft Hall*, appeared on shed on 11 July 1965 bearing a Bristol (Barrow Road) shedplate. How this engine got there or what it was doing is a mystery that my extensive research has failed to solve. The likeliest assumption is that she had been booked to work a freight from Shrewsbury to Chester over former GWR metals, and had worked through to Birkenhead. She was among the final withdrawals of the class the following December.

the line's lighter duties. Trains on 'the Branch' – the Evercreech Junction to Highbridge and Burnham-on-Sea line – were never more than light-weight, so the pannier tank seen (Plate 14) in charge of the 16.00 train from Highbridge to Evercreech Junction on 20 February 1965 is perfectly adequate, from the standpoint of tractive effort if not aesthetics. Nevertheless, S&D men insisted that the GWR engines badly battered the track across the Somerset levels.

However, factual details of tractive effort, or arguments about the advantages of taper boilers were often merely cloaks to hide the fierce loyalty we felt to *our* railway, a loyalty that allowed us to enjoy the sight of a 'foreigner' on *our* line, so that we could extol the invincible superiority of our locomotives, our railway . . .

3T37
3T37
8Z09

13 Birkenhead

'I, the returning officer for the Birkenhead Parliamentary Constituency, do hereby record the number of votes cast for each of the candidates at the election ... Adley, Robert James 15,438 ...'

Thus ended a glorious spell of my life, not as a politician but as a railway enthusiast!

It is said that God helps those who help themselves. Luck in life is important; yet even luck depends on the ability to recognise and grasp the opportunities thrust in your path. If I am to ascribe thanks to those whose advice sent me winging on my way to be the Conservative Candidate for Birkenhead in the 1966 General Election, then mention must be made of Sir Paul Bryan, who as plain 'Mr' in 1965, was Vice-Chairman of the Conservative Party, in charge of candidates. His suggestion to an aspiring candidate being interviewed at Central Office, was sound. 'Fighting your first seat, you are bound to make mistakes – so go somewhere where it doesn't matter too much. Choose somewhere two hundred miles from home – then you can learn something useful about an unfamiliar part of the country.'

With my target thus defined, a 'safe' Labour seat about two hundred miles from Sunningdale, I sought to narrow the target area. An examination of my 'steam map' clearly indicated the North-West, and I waited for Central Office to send the list of constituencies seeking candidates. A few days after I'd had my name added to the 'approved list', I received my first communication. 'Candidates are invited to indicate to which constituencies they would like their name to be forwarded for consideration by the Constituency Selection Committee'.

Not suffering from delusions of grandeur, I did not submit my name for a safe Tory seat, but, recalling Paul Bryan's words about mistakes, decided I wanted a seat with a Labour majority, but of four figure rather than five figure variety! There it was – Salford West! I ticked the box alongside, sent back the form and waited.

PLATE 56. This was taken on 8 August 1965 at Birkenhead MPD, yet a quick glance at this LMS line-up might date it a generation earlier. Even on a fine day the atmosphere could be choking. 'Fairburn' 2–6–4T No. 42121 poses alongside 'Black 5' Variant No. 44697 and '8F' No. 48425.

To my amazement I received a letter, a few days later, from the agent to the Salford West Conservative Association inviting me to appear before the selection committee for interview. As Sales Director of the May Fair hotel, I was master of my movements, so my arrangements were easily made. I 'passed' the first interview, and was invited to appear on their short list for final selection. Alas the date chosen coincided with a sales visit to Scandinavia! However, on assurance from the agent that the selection committee had been 'most impressed' with my appearance before them, I laid elaborate and expensive plans to enable me to fulfil my appearance on their short list. Flying back from Stockholm, I hired a car at Heathrow and headed north for Salford. Running out of petrol on the way, I reached the Conservative office at Salford West with only five minutes to spare.

'Oh, hello Mr Adley', they said. 'Nice of you to come. We tried to reach you, to tell you we've decided to re-select Albert Clark, as the next election will probably not be long delayed. However, since you're here, would you like to address the Selection Committee anyway?'

As I stormed back to London and took the plane to Stockholm, to resume my Scandinavian sales tour, I felt like terminating my political 'career' there and then.

On returning home from the sales tour, I stared long and hard at the next list of constituencies seeking candidates. Birkenhead fitted my criteria of a 'north-west labour seat near steam', so I returned the form, passed the initial 'inspection' and was shortlisted. I was selected as Prospective Conservative Parliamentary Candidate for Birkenhead on Thursday 22 April 1965, on the basis of the first political speech I had ever made in my life. Thus was I enabled to go 'in search of steam' in the North-West!

My, perhaps unusual, combined dedication to politics and railways no doubt mystified some of my new-found friends in Birkenhead. The *Birkenhead News* gave my adoption good coverage: in the leading article the Editor issued a friendly warning: 'Mr Adley need have no illusions about Birkenhead. He is coming to a tough constituency. . . .' Maybe, but never before or since have Jane and I made more friends more quickly, included among whom were several with railway connections.

The men at Birkenhead MPD were taken aback to find the Conservative Candidate amongst them. Inquisitive they may have been, but hostile they were not. Soon I was being invited into railway homes, and sharing reminiscences with men who had given a lifetime to the Cheshire Lines Committee (CLC), Great Western, London and North Western and latterly the LMS and British Railways. Inevitably my railway enthusiasm soon reached the ears of the local press, and the *Birkenhead News* carried a shot

Fig. 13 (page 127) and Fig. 14 (Page 128).
Notwithstanding the retention of the diesel depot at the Mollington Street site, a glance at the OS maps of the immediate vicinity of the engine shed, dated 1963 and 1973 well illustrates the changes in the railway scene wrought by the ending of the steam era. The reduction in the number of roads into the shed: the elimination of the coal yard: and the removal of the complex of lines, sidings, signals, signal-box, water-columns, turntable: all serve to reduce the importance of, and interest in the railway landscape.

Whilst the railway has contracted, the roads have expanded. The construction of the new viaduct (top right) has coincided with the destruction of many homes in the Hind Street/Thomas Street/Jackson Street area.

Notice the replacement of the Hind Street Methodist Church by the Church of Jesus Christ of Latter Day Saints.

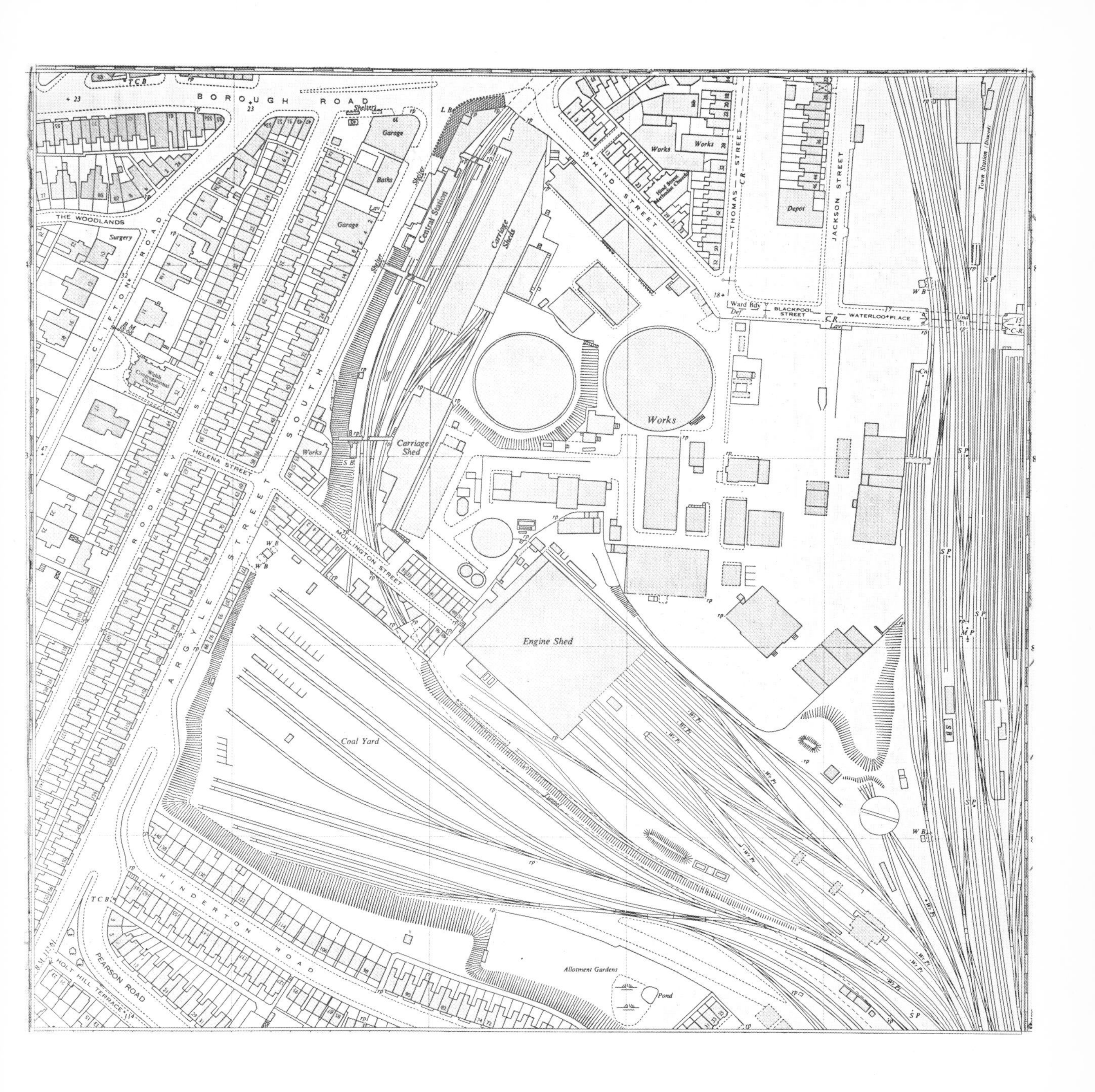
BOROUGH ROAD
Shelters
Garage
Baths
Garage
Shelter
Central Station
Carriage Sheds
HIND STREET
Works
Works
Hind Street Methodist Church
THOMAS STREET
Depot
JACKSON STREET
Town Station (Disused)
THE WOODLANDS
Surgery
CLIFTON ROAD
Welsh Congregational Church
SOUTH
RODNEY STREET
HELENA STREET
ARGYLE STREET
Works
Carriage Shed
Works
Ward Bdy
BLACKPOOL STREET
WATERLOO PLACE
MOLLINGTON STREET
Engine Shed
Coal Yard
Tunnel
HINDERTON ROAD
PEARSON ROAD
HOLT HILL TERRACE
Allotment Gardens
Pond

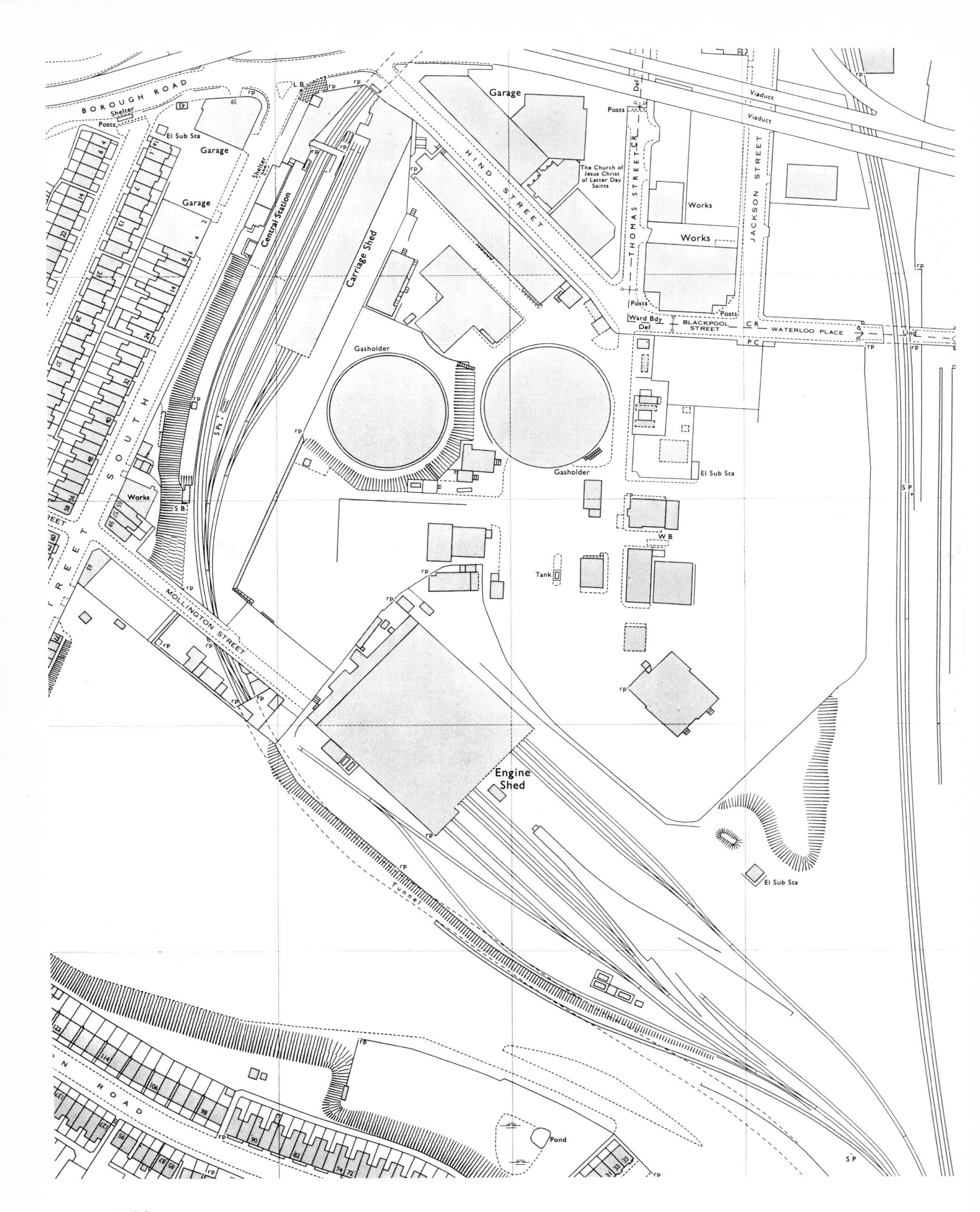
BOROUGH ROAD
Shelter
Posts
El Sub Sta
Garage
Garage
Central Station
Carriage Shed
Garage
HIND STREET
The Church of Jesus Christ of Latter Day Saints
THOMAS STREET
JACKSON STREET
Works
Works
Viaduct
Viaduct
Posts
Ward Bdy
BLACKPOOL STREET
WATERLOO PLACE
Gasholder
Gasholder
El Sub Sta
SOUTH
STREET
Works
MOLLINGTON STREET
Tank
W B
Engine Shed
Tunnel
El Sub Sta
ROAD
Pond

PLATE 57. Birkenhead shed was essentially a place for man's work; heavy and frequent freight services dominated the railway environment. The latter days of steam seem best illustrated by Stanier 8Fs, such as 48505 and 48253, resting on shed on 7 May 1966, my last visit to a steam Birkenhead.

of me in the cab of 9F 2–10–0 No. 92079 in what the caption described as 'Hinderton Road coal sidings'; these were alongside Birkenhead MPD, the entrance to which was in Mollington Street.

A look at the Ordnance Survey maps of the shed and adjacent coal yard, dated 1963 and 1973, illustrates all too well the justification of my concern, reported in the local press in 1965, that the demise of steam would have a bad effect on employment in Birkenhead. I am glad, however, that the diesel depot at Mollington Street still provides employment for some of my former prospective constituents.

Birkenhead has a fascinating railway history. The old BJ – Birkenhead Joint – was operated by the LNWR and the GWR. The former was merged into the LMS at the 1923 grouping, yet even with nationalisation the shed at Birkenhead retained its twin ex LMS/GWR status. The northern outpost of the GWR main line from Paddington, Great Western engines survived at Mollington Street until 1963, notwithstanding the fact that the depot came under BR's Midland Region.

On 31 December 1947, there were forty-two GWR engines allocated to the shed, of which eighteen were 0–6–0 pannier tanks, reflecting the extent of freight, dock and shunting work in the area.

By the time that I paid my first photographic visit to the shed, on 8 August 1965, GWR engines were but a memory. In all my numerous visits to the shed between then and its closure, only once did I photograph a Swindon engine, on 11 November 1965. The shot of 7924 *Thornycroft Hall* alongside LMS 'Jinty' 0–6–0T 47627 (Plate 55) and BR Standard '9F' 2–10–0 92157 happily bridges Birkenhead's railway generation gap.

Birkenhead shed was the quintessence of a steam engine shed. Built jointly by GWR and LNWR in 1878, it was a smoky, crumbling relic of the Victorian era by the mid '60s. Inside the reduced-size shed which followed the cutting back of the former LNWR side in 1951, the atmosphere was chokingly acrid. The men's facilities were spartan at best, and only dedicated railwaymen could remain cheerful whilst working in such foul conditions. It was not much better outside. Even on a clear summer day there was a constant haze, and in the absence of wind the air was thick – but marvellous!

The smell of the steam railway was everywhere. As late as the winter of 1965/66 there was plenty of work for the shed's allocation of LMS and BR/STD engines. The 9Fs worked the heavy John Summers ore trains, Crabs handled freight work, 3F' 0–6–0Ts handled local freight and shunting duties and LMS 2–6–4Ts handled the passenger trains. GWR through trains from Birkenhead to Paddington were but a memory.

Trains ran from Birkenhead Woodside terminus to Chester, along the route of the Chester & Birkenhead Railway, opened as long ago as 23 September 1840. An occasional Crab could still be seen on passenger workings into Woodside.

The Crabs had a long reign at Birkenhead. These fine engines (see Chapter 4) had an outstanding record of reliability, an attribute which rarely cast them in a leading role, yet which ensured their position as favoured engines in the sheds where they were allocated. The last five active survivors of the class ended their days here: no. 42942 working the 05.40 Bidston Yard to Dee Marsh pick-up freight on the morning of 14 January 1967 was the Crab's revenue-earning swansong. Plate 19 shows detail of the motion of one of Birkenhead's Crabs.

Woodside Station epitomised the town's status vis-à-vis Liverpool. Opened in 1878, the red-brick Venetian Gothic station always seemed ashamed of itself! The main entrance as built was not used, access to the station being through an inconspicuous side-entrance.

I still have a copy of the Western Region timetable (price one shilling) valid at the date of my adoption as Birkenhead Prospective Conservative Candidate. Seven trains ran from Woodside to Paddington on weekdays, including the 20.55 sleeping car service which arrived in London at 05.10. In today's composite timetable (1980/81 price £2.50) the name 'Birkenhead Woodside' no longer exists. There is 'Central' on the cross-Mersey service from Moorfields and Liverpool Lime Street to Rock Ferry. Hamilton Square is on this line, and on the Moorfields to New Brighton and Moorfields to West Kirby lines, along with Birkenhead Park and Birkenhead North. In railway terms, the town has sunk to the status of suburban Liverpool.

My candidature at Birkenhead enabled me to search for steam around Liverpool, too. The Edge Hill carriage sidings presented an extraordinary sight. By July 1965 the overhead catenaries carrying the power-lines for the electrification of the West Coast Main Line (WCML) to Manchester and Liverpool were in place, but pending the introduction of full electric services, many local freight and shunting duties were still performed by steam. Edge Hill, Bank Hall and Speke Junction were three of the last surviving steam sheds. An LMS '3F' 0–6–0T, with its vintage appearance, shunting 'under the wires' totally personified the changing railway scene represented by the move from steam to diesel and electric motive power.

As with all shed visits in the declining years of steam, engines presented an ever-more decrepit sight. Nothing caused much surprise any longer, not even the sight of a Black 5 at Edge Hill minus its middle pair of driving-wheels! As this shot was taken only ten miles from the setting for *Rocket's* triumph at Rainhill in 1830, I suppose one should accept with nonchalance the Liverpudlian's view of the miraculous qualities of their city – for creating a 'Black 5' with 4–4–0 wheel arrangement!

During that 1966 election campaign I insisted on keeping Sundays free from politics. On the middle Sunday of the campaign I set out on an expedition in search of steam accompanied by Bert Price, my host for the duration of the election. We headed for North Wales, hoping somewhat forlornly to glimpse the last remains of GWR steam. Approaching Wrexham we located the shed at Croes Newydd. The atmosphere was of death before burial. Towards the end of a glorious afternoon, 'twixt winter and spring, we passed silently from engine to engine paying our respects (see Plate 15). The roundhouse was deserted. Inside, dead pannier tanks and 56xx 0–6–2 T's stood in silent dignity. Outside, the fires had not yet died on one or two engines. . . .

Back in Birkenhead we reminisced that Sunday evening. My love of the steam engine had ensured a day of total relaxation from politics: a blessing in itself. There remained only a few more months of steam-working even for Birkenhead.

With steam went status for Birkenhead. Even the sight of a Stanier 2–6–4T storming up past the shed and under Green Lane bridge is just a fantasy from yesteryear. . . .

PLATE 58. Fairburn 2–6–4T No. 42156 pounds up the grade and past the motive power depot, with empty stock from Birkenhead Woodside, on 9 January 1966. *(Following pages.)*

1V23
61327
70024

14 'Twixt Highbridge and Bath

If you are puzzled on reading this chapter title, it is not surprising! Bath and Highbridge were both termini of the Somerset and Dorset Joint Railway: yet the obvious way to travel between the two places was on the GWR line via Bristol. Being unable to spare the space to devote a chapter to Bristol alone, and having S&D photographs taken mainly at Highbridge, I was faced with a dilemma. Various titles were tentatively attached to this chapter: 'Around Bristol' survived for a couple of weeks: 'Somerset & Dorset' seemed likely to contravene the Trades Descriptions Act: 'Steam in the West' was both boring and inaccurate; and 'GWR & S&D' was ungainly. I even tried 'Avon' as the new county stretches from Bath to Weston-super-Mare, but I wanted to include shots at Highbridge and at Edithmead, both of which, doubtless to the pleasure of the inhabitants, remain in Somerset. So 'Twixt Highbridge and Bath' prevailed.

Notwithstanding the title, Bristol claims pride of place as the Railway Capital of the West. Whilst Crewe, Swindon, Doncaster and Eastleigh might claim, with justification, to be Railway Towns in the sense that the railway dominated their social and commercial environment, few great cities, as opposed to mere towns, played a more significant role than Bristol in our railway's development. The city is Brunel's spiritual home; and Brunel must rank as the outstanding Victorian railway engineering genius. He traced all his subsequent success to his victory in the Clifton Bridge competition of 1829. Brunel always remembered this link with the great city of Bristol. In 1833, at the age of 26, he was appointed engineer to the Great Western Railway Company, then about to begin building its main line from Bristol to London. Bristol Temple Meads station still retains an atmosphere of importance, of grandiose activity.

Borne of Brunel's 'train shed' (the phrase itself evokes the pioneering spirit), the station was one of the great meeting places of GWR & LMS. Just as the Almondsbury interchange of M4 and

PLATE 59. Were it not for the cosmopolitan nature of the railway world around Bristol, this shot of LNER 'B1' 4–6–0 No. 61327, from Canklow, could have been included in Chapter 12. This locomotive, seen at Barrow Road on 27 June 1964, had worked right through from its Yorkshire home, on a train from Leeds to Bristol.

Fig. XV. Brunel's 'Train Shed' at Bristol Temple Meads: now there is a scheme to restore it to its former glory.

M5 motorways is one of our great transportation interchanges, so Temple Meads remains what it was; a pre-eminent crossover point for rail services between the South-West and Midlands and North-West; south coast to South Wales; and the London-Bristol main line. It is worth adding that the new station north of the city on the London/South Wales line is called Bristol Parkway.

There were three important steam MPDs in Bristol: Bath Road, St Philips Marsh and Barrow Road. The latter was the LMS shed, the first two belonged to the GWR. Whilst Bath Road, alongside Temple Meads station is still in use as a diesel depot, the two other sheds have gone. St Philips Marsh was demolished long ago, but not before I was enabled to record its existence in colour. The site was sold to the City of Bristol, and has been developed as a modern market/warehousing complex. There is now no trace of its dignified past.

St Philips Marsh was closed on 15 June 1964, and its allocation of GWR engines transferred across the city to Barrow Road. For a few hectic weeks Barrow Road enjoyed the experience of an over-allocation of steam, an unforgettable interlude in steam's declining years.

PLATE 60. No. 6927 'Lilford Hall' adds a touch of life to a grim winter's day in January 1963, as she prepares to depart from Bristol Temple Meads with the 10.50 train to Salisbury.

To return to the site now, seventeen years later, is a sombre experience. The motive power depot was demolished and the whole site cleared years ago. But its future still uncertain, it

PLATE 61. One for my family album: nephews Nicholas and Timothy Pople pose, not without misgivings, alongside the 6′ 0″ driving-wheels of GWR 'Hall' class 4–6–0 No. 6916 *Misterton Hall*, at Barrow Road, Bristol, in June 1964.

remains a haunted land of railway long-ago to those who knew its glory. The great shed and distinctive chimneys have gone, replaced by willow, budleia, couch and fescue. Standing amongst the weeds, looking under the arches of Midland Road Bridge, the memories flood back, and I was able to capture the atmosphere of Barrow Road before it died.

The former Midland Railway line into Bristol ran alongside the shed, and the running lines are still used as storage sidings for the nearby wagon maintenance depot. Indeed the site is still classed by BR as 'operational', with the possibility of use as a rail-served terminal for a refuse disposal plant. To me, with my memories, that is like converting a cathedral into a supermarket . . .

The revival of plans to convert Brunel's 'train shed' into a railway museum, is another belated attempt to ensure for Bristol its rightful place as a centre of commemoration of the steam era. One day, when I am older, I may tell the true tale of the obstruction, objection and even ridicule that confronted Richard Goold-Adams, Ewan Corlett and myself when we started the project to return Brunel's great iron steam ship, the SS *Great Britain* from the Falkland Islands to Bristol. The attitude of those involved contrasted sadly with that of the great railway pioneers.

Nowhere was the spirit of the railway pioneers personified

PLATE 62. Traditional inhabitant: LMS '4F' 0–6–0 No. 44466 simmers quietly at Barrow Road, Bristol, on 27 June 1964. For a few hectic weeks following the closure of St Philips Marsh, the shed at Barrow Road was overcrowded and intensely busy; it was marvellous! *(Above.)*

PLATE 63. 'But some there be who have no memorial . . .' Locomotives withdrawn from Barrow Road and dumped on the sidings alongside the depot, await their journey to the breakers yard – January 1963.

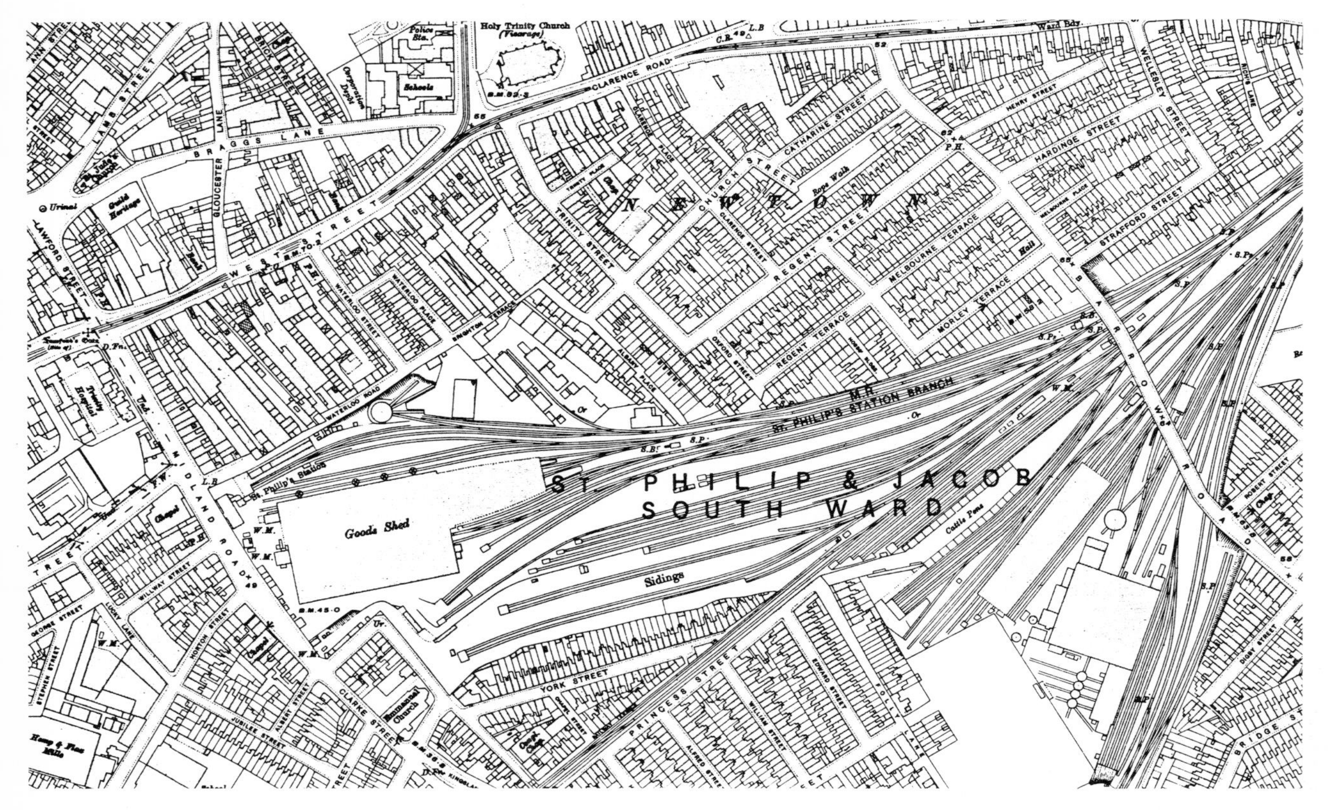

more than in the building and operating of the Somerset & Dorset, about which probably more words have been written, and more photographs taken per mile of track than on any line on earth. Indeed, the fierce rivalry between the old Midland Railway and the Great Western brought about the role of through-route enjoyed by the S&D from Bristol to English Channels, which came from the alliance between the Midland and the LSWR aimed at thwarting the ambitions of the GWR.

By the time my camera was first aimed at steam at Highbridge, the GWR takeover of the S&D – or, to be more accurate, the Western Region takeover – was already a *fait accompli*, and the line's future was finite. Burnham-on-Sea, home of Jane's family, had already lost its rail service, regular passenger trains being extinguished on 29 October 1951. Although an unrepentantly fanatical admirer of the GWR, even I regret the way that a century of rivalry followed by nationalisation, didn't dim the gleam in the eye of GWR when they spotted the chance to do down an old rival. This historic hostility hastened the demise of the S&D, as it did to virtually all the former LSWR lines west of Exeter; indeed it almost brought about the closure of the Salisbury to Exeter line, succeeding in reducing this once-proud trunk line to a mere twig. But that is another story. Whilst the line

Fig. 16 (page 140), Fig. 17 and Fig. 18 (page 141, top and below).

A glance at the OS maps of the Barrow Road area in Bristol shows clearly how the area has declined as a centre of railway activity. The 25″ map, revised 1918 indicates not only the complex of buildings comprising the engine shed, but also the adjacent St Philips Station and Goods Shed – not to be confused with St Philips Marsh engine shed a mile or so away. Indeed almost the only constant feature on the three maps of the area of Barrow Road, dated 1918, 1951 and 1977 is Barrow Road itself.

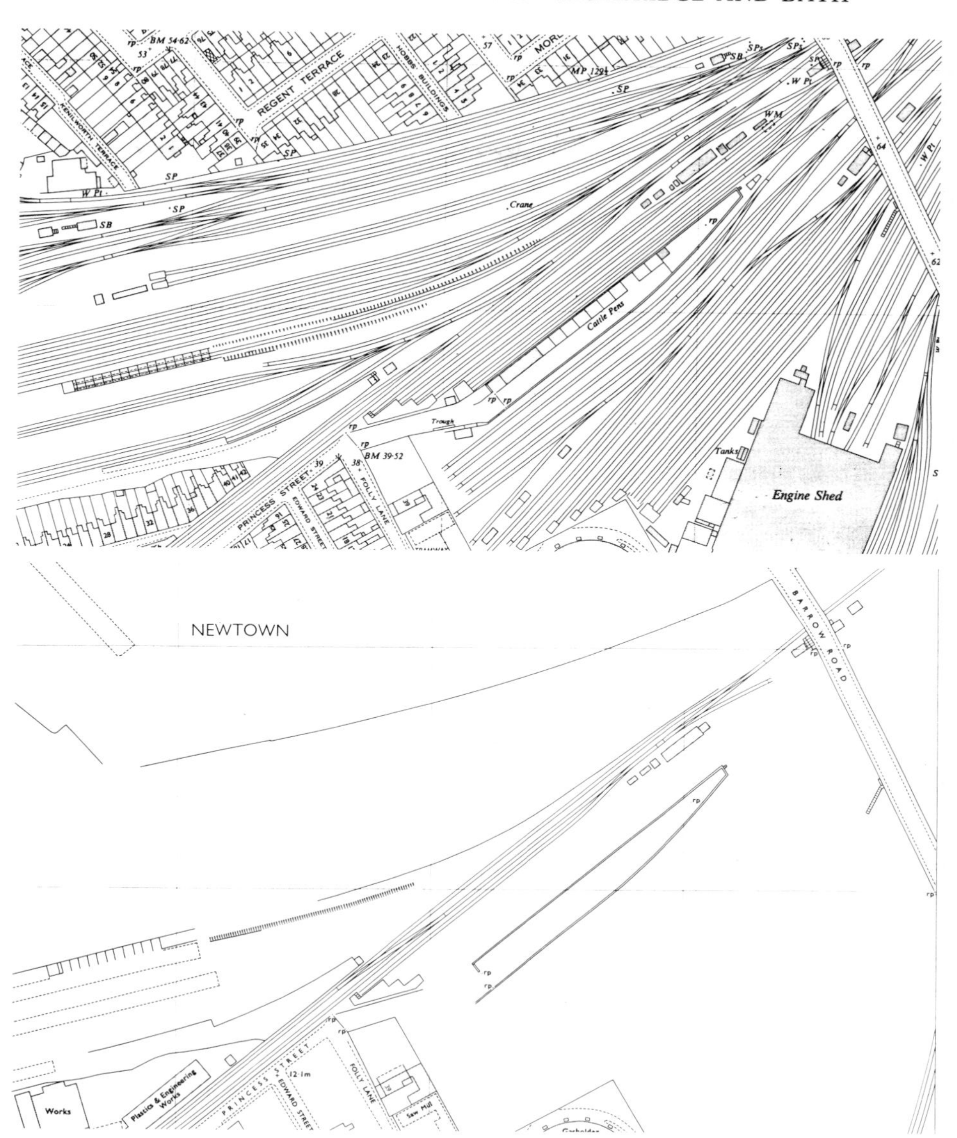

April 1980 I took my first pictures at Barrow Road since 1964. (The gasometers are located in Fig 16 in the blank oblong in the bottom right hand corner of the map: They are just visible in the two following maps.) The contrast on the ground was as dramatic as indicated on the maps, particularly clearly shown by those on the same scale (fig. 17 and fig. 18). As I aimed my camera at the gasometers my attention was drawn to a television camera crew shooting the same subject. At the home of my mother-in-law in Sneyd Park, Bristol, that evening I was to learn the reason for their interest in those gasholders.
A pair of nesting kestrels had attracted the television men, whose pictures appeared on BBC TV 'Points West' that evening. Indeed those kestrels have extended the life of the Barrow Road Gasometers, which reprieved by South West Gas, still stand (April 1981) to retain a link with Barrow Road times past.

from Highbridge to Evercreech Junction had long since become known on the S&D as 'the Branch'; the GWR line which it bisected at the notorious flat crossing at Highbridge was the GWR trunk route from Bristol to Exeter, one of the first important railways to be built. Before the construction of the 'Berks & Hants' line of the GWR, it was the route for all Great Western through services between London and the South-West. Today the route remains a vital rail link, being the South-West to North main line. In the latest timetable, some Paddington-Plymouth trains are booked via Bristol, passing through Temple Meads without stopping.

As steam on the GWR approached its final months, the steady

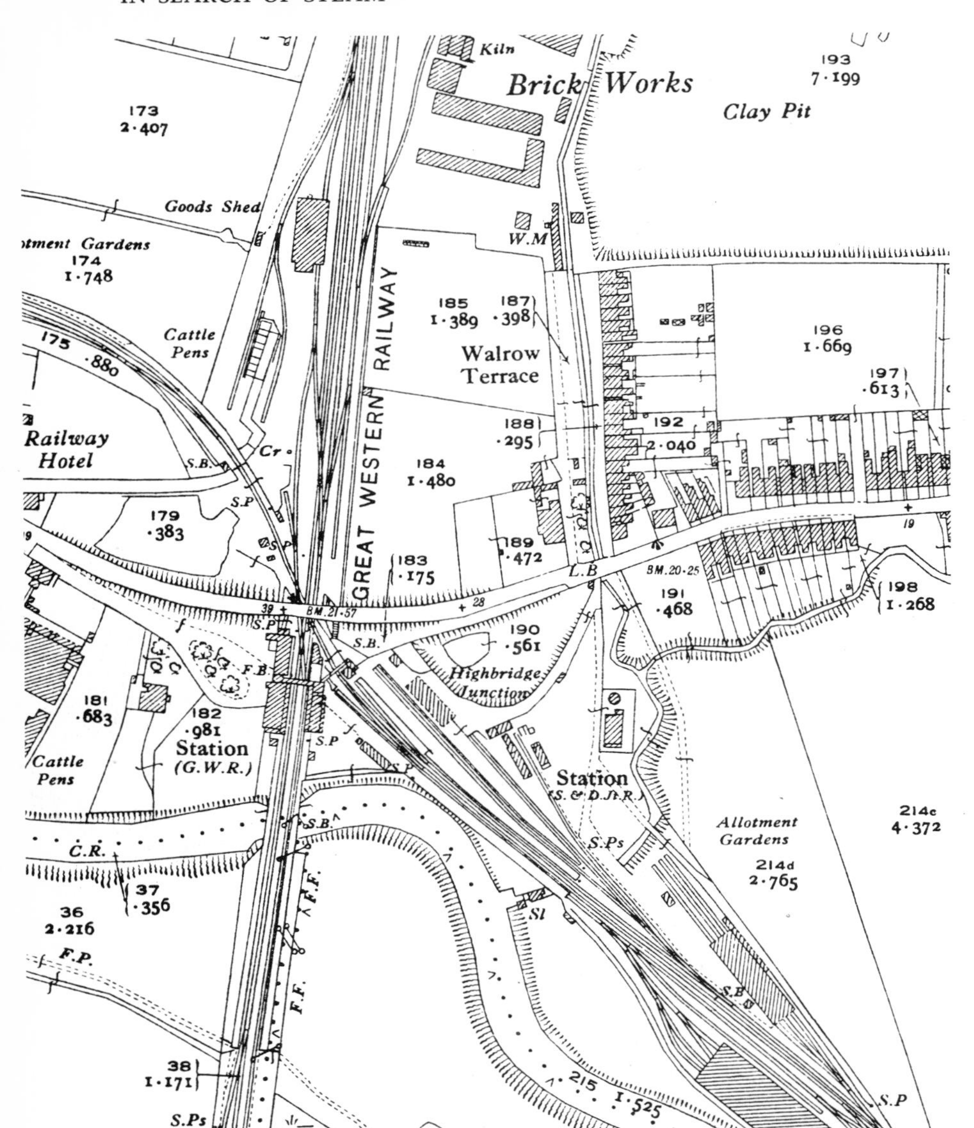

Fig. 19 (page 142) and Fig. 20 (page 143).
Few railway locations in Britain have undergone such a transformation as has Highbridge. The OS map dated 1930 illustrates not only the extent of the works and track, but shows the well-defined separate existence of the S & DJR and GWR stations. The Burnham-on-Sea line can be seen curving away to the left, and four signal boxes are still in use.
The S & D station had five platforms, but after the line's relegation to branch status, and accentuated by the 1951 closure of the Burnham line to regular passenger services, it became a ghostly near-deserted place, which closed to goods traffic on 2 November 1964 and finally to passengers on 7 March 1966. On the OS map revised to 1970 the remnants of the S & D remain, in the shape of the connection to Bason Bridge; it was merely an extended siding. Even this line, serving the milk factory, closed on 2 October 1972, and Highbridge today is merely a glorified halt on the GWR line between Bristol and Taunton.

stream of holiday rail traffic between the industrial Midlands and the North, and the resorts of Devon and Cornwall, moved from the hands of steam to diesel. On my weekend visits to Burnham I would sometimes forsake the S&D for the GWR main line. By the height of the summer holiday season in 1963 I recall three hours spent under a glorious sun in Somerset on the afternoon of 27 July, yet my reward in all this time was one steam-hauled express. At the time it seemed an ill-judged expenditure of time, yet the picture of the train concerned, the 08.35 (SO) from Ilfracombe to Manchester Exchange, seen in Plate 64, seems in retrospect to justify the investment of those three hours.

Whilst Highbridge was the unenthusiastic meeting-place of

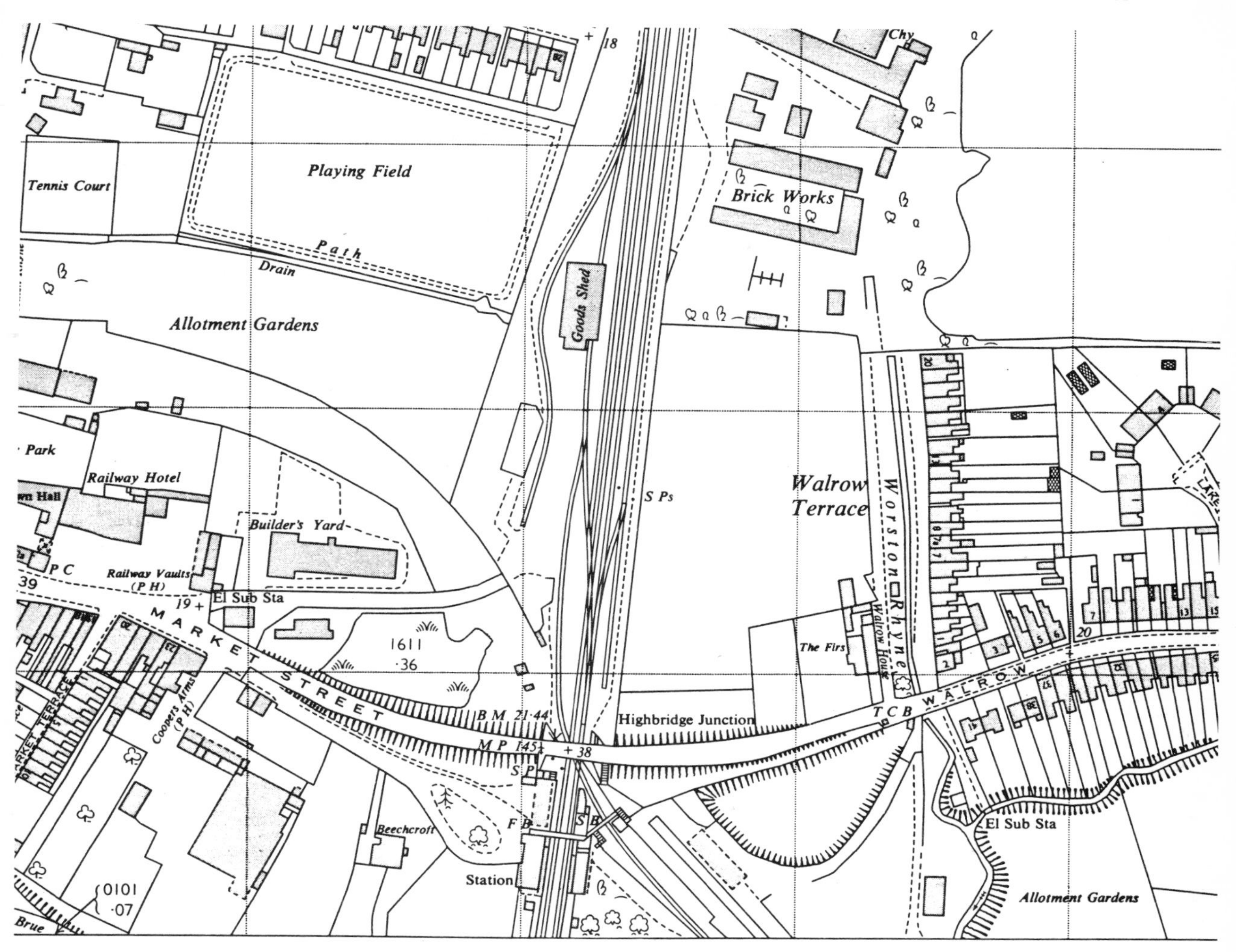

GWR and S&D, Bristol was the natural rendezvous of GWR and LMS. With pre-grouping rivalries forgiven if not entirely forgotten, these two great companies had long since learned to co-operate to their mutual benefit. Between Bristol and Highbridge lies Weston-super-Mare. Now reduced to single-line status, the Weston loop once echoed to the bark of Castles, Halls and Granges on 'Summer Saturday' trains. Although originally the terminus of a short branch off the Bristol and Exeter main line, the line was continued westwards in 1882 to form a through-loop, rejoining the main line at Uphill Junction, three miles west of the original branch junction.

A diminutive stone engine shed, spanning a single road, had been built in 1861 to serve the small branch terminus of what became known in later years as Locking Road station. Although the shed was officially closed in August 1960, it continued to offer hospitality to the ever-diminishing number of steam-hauled

PLATE 64. The sole fruit of a three-hour stint beside the track at Edith Mead, near Brent Knoll, Somerset, on 27 July 1963. Steam-hauled passenger expresses were already a rarity on BR Western Region when I captured this shot of No. 4993 *Dalton Hall* passing Edith Mead with the 08.35 (SO) Ilfracombe to Manchester Exchange train. Manchester Exchange has now vanished from the railway timetables. *(Opposite.)*

PLATE 65. Summer excursion scene from yesteryear: engines which have hauled trains to Weston-super-Mare, bask in the sunshine after servicing at Weston's diminutive depot. In the foreground is Collett 4–6–0 No. 6852 *Headbourne Grange*. *(Above.)*

excursions. On a splendid June day in 1963, I was fortunate to see 7034 *Ince Castle* and 6852 *Headbourne Grange* in immaculate condition resting on shed. The picture of the latter engine is amongst my more cherished shots, because no Grange has been preserved and colour photographs of Weston MPD are not exactly common!

My time spent at Bath with camera was limited, but at least I was able to record one of the distinctive Derby-built S&D '7F' 2–8–0s in its natural environment. Yet my final and abiding memory of steam 'twixt Highbridge and Bath' concerns the Wilts United Dairies milk factory siding at Bason Bridge, one mile south-east of Highbridge on the S&D. Here, as at innumerable rural villages in the steam era, was played out the activity of 'shunting the milk'.

The procedure at Bason Bridge was as follows. The milk company would fill rail tankers with milk. The tankers would be collected by a locomotive that had worked 'light-engine' along 'the Branch' from Highbridge. Having collected the milk tanks, the designated engine brought them up from Bason Bridge to Highbridge S&D, traversed the flat crossing of the GWR and by means of the stump of the erstwhile Burnham branch, then propelled the tanks from the S&D to the GWR down line at

Highbridge. From the down line they were then shunted onto the up line, and finally placed in the holding loop immediately to the south of the up platform at Highbridge GWR station ready for an eventual journey to Kensington. If it sounds complicated, it was!

On 18 August 1963 I was invited to ride on the footplate of the Collett 0–6–0 on the Bason Bridge milk duty. We clanked merrily across the peaceful Somerset flatlands, the line running alongside the River Brue. Cows chewed the cud, bees buzzed, a small boy was fishing in the river near Newbridge Farm. I swear that, if I could choose a moment in my life to be embalmed in time, then that afternoon on the footplate of 2204 would be on the shortlist.

All trace of that way of life, indeed all sign of life on the S&D at

PLATE 66. Summer in Somerset, as 'Collett' 0–6–0 No. 2204 collects tanks from the milk factory at Bason Bridge. They will be taken along the S&D to Highbridge. This photograph was taken on 18 August 1963. *(Below.)*

PLATE 67. My attachment to Somerset is my excuse for including this picture of my late father-in-law, Wilfrid Pople. The location is the railway bridge that carried the S&D branch from Edington Junction to Bridgwater North, over the road in the village of Bawdrip. The site today, apart from the bridge, denies the very existence of the railway, still visible here in October 1963. *(Above.)*

Bason Bridge has gone the way of all flesh. No longer can one glimpse plumes of white smoke from Mark Causeway. South of Bason Bridge the name of Edington Junction has gone, too. From here ran the line to Bridgwater via Cossington – more a twig than a branch. It was opened in 1890 and closed in 1954. The last photograph I took of my late father-in-law was on the bridge that carried the S&D Bridgwater branch over the lane at Bawdrip by the church, before the line crossed King's Sedgemoor Drain, ran under the A39 Bridgwater–Glastonbury road, then over the GWR line and the A38 before terminating in Bridgwater. If, as I did on 7 June 1980, you visit Bason Bridge or stand on the remains of the platform at Edington Junction, or fight your way through the brambles on the bridge at Bawdrip you will feel the unquenchable spirit of S&D; gone but not forgotten. . . .

15 Barry

A flamboyant scrapmetal man in South Wales is an unlikely aspirant to the title 'Last Hero of Steam'. Dai Woodham, owner of Woodhams, Barry, could – would, were he not so modest – indeed claim such a title. Pure chance has seen created in Barry's decaying docklands a Mecca for the steam enthusiast, to which hundreds of thousands have flocked these past few years.

1981 sees the twentieth anniversary of the arrival at Woodhams Yard, of Barry's longest-serving resident, GWR 2–6–2T No. 5552, withdrawn in October 1960. Since LMS 4F 0–6–0 No. 43924 left Barry in 1968, well over 100 engines have departed for preservation and re-birth. The quirk of fate which caused Dai Woodham to delay their date with destiny, or the cutter's torch to be more prosaic, was a circumstance which has totally transformed Britain's railway preservation scene.

The tale has been told elsewhere; it is not for me to chronicle the exploits of the countless thousands of the intrepid, the brave, the fanatical and the foolhardy – adjectives abound to describe the men and women who have joined the Barry Crusade.

This chapter is merely Valete – Steam. Prose cannot capture the atmosphere of Barry, so a final tribute to the steam engine is best conveyed in verse.

My humble tribute opposite, is to the thousands whose enthusiasm, energy, initiative and determination, have rescued more than a hundred derelict steam engines from Dai Woodham's scrapyard in Barry, South Wales

As Mecca to the faithful
Or to cricketers its Lords
Or Wembley to the seething mass
of loutish soccer hordes:
A place to come to pay respect
Enthusiasts must not neglect
Yet stop. You cannot comprehend
how Barry is the very end
of life: not sport, nor e'en the spirit.

To see ones faithful friends in death
Unburied: left to rot – the breath
long gone. The skin: the bones
Stripped, scarred, silent, alone,
Forgotten. Then, the years now tolled,
The graveyard visited. One bold
Adventurous determined dream
Vowed to return: to steam.

Again: removing from this tomb,
This place of all-embracing gloom
Reincarnation to the womb
By man created resurrection.

Slowly the years have yielded the dead
Unimagined the hope that has come in its stead.
It's as though we have managed to bring back to life
Our deeply-mourned friends who were put to the knife.
If not murdered – then foully scarred: the cut cord.
Yet fate has decreed that some be restored.

The Barry Obituary – last to be written of steam?
They can no more extinguish these creatures than banish a dream!
The race is now on: the survivors await in a herd
Their rescuers. Which one will next be preferred?
Some yet will be saved. Another miraculous chapter
Remains to be written. The future in steam to enrapture . . .

They will not all make it: but nothing can dim
The song that's been sung: a magnificent hymn
The spirit of Bulleid, of Stanier, Churchward
of Maunsell and Riddles, of Collett: all nurtured
On faith that created man's worthiest dream
The brotherhood known by the single word – steam!

PLATES 68. Unstinted admiration for the preserved railways cannot mask the greater nostalgia epitomised by these engines at Barry in January 1980. Their very survival is astonishing: the rescue and restoration of many is little short of miraculous. As chairman of Barry Rescue I hope to help save some more of the engines.
Stanier 8F 2–8–0 No. 48305, solemnly scrutinised by Dai Woodham and my wife Jane. *(Top left)*.
Maunsell U 2–6–0 No. 31625 has survived, and went to the Mid-Hants Railway in March 1980. *(Bottom left)*.
Dai and Jane ponder the fate of an unrebuilt Bulleid West Country Pacific. *(Top right)*.
The future for this GWR 2–8–0T is bleak. *(Bottom right)*.

34007

ABBREVIATIONS

AWS	Automatic Warning System
BR/STD	British Rail Standard Classes of Steam Engine
CLC	Cheshire Lines Committee
CME	Chief Mechanical Engineer
d.m.u.	Diesel Multiple Unit
ECML	East Coast Main Line
ECS	Empty Carriage Stock
GNR	Great Northern Railway
GWR	Great Western Railway
LBSCR	London Brighton & South Coast Railway
LCD	London Chatham & Dover Railway
LMS	London Midland & Scottish Railway
LNER	London & North Eastern Railway
LNWR	London & North Western Railway
LSWR	London & South Western Railway
L&YR	Lancashire & Yorkshire Railway
Mogul	An engine of 2–6–0 wheel-arrangement
MPD	Motive Power Depot
MR	Midland Railway
MT	Mixed Traffic (eg 4MT, 5MT) Engine Classification
Pacific	An engine of 4–6–2 wheel-arrangement
6–PAN 6–PUL	6-Car Electric multiple units including a Pantry or Pullman car
PT	Pannier Tank
RCTS	Railway Correspondence & Travel Society
S&DJR	Somerset & Dorset Joint Railway
SECR	South Eastern & Chatham Railway
SER	South Eastern Railway
SLS	Stephenson Locomotive Society
SR	Southern Railway
T	(after wheel-arrangement) Denotes tank engine
WCML	West Coast Main Line
WD	War Department (engines built under Government auspices during World War 2, subsequently sold into BR stock and classified WD)
WLER	West London Extension Railway
2–BIL	Two-car electric multiple unit
4–LAV	Four-car electric multiple unit containing one lavatory

Index

Included in this index are page references shown in bold type to the colour and black and white photographs: to the Ordnance Survey maps: to locomotives where a textual reference complements one of the photographs: and to people and places relevant to the photography and narrative.

It is *not* intended to index every place, person, event or locomotive mentioned in the text. The post-grouping 'Big Four' are not indexed, to avoid repetition: nor are the British Rail regions, or BR itself.

*See also under 'Motive Power Depots'.